AD-Vol. 36

FATIGUE AND FRACTURE OF AEROSPACE STRUCTURAL MATERIALS

presented at
The 1993 ASME Winter Annual Meeting
New Orleans, Louisiana
November 28–December 3, 1993

sponsored by
The Aerospace Division, ASME

edited by
Arvind Nagar
Wright Laboratory

An-Yu Kuo
Structural Integrity Associates

THE AMERICAN SOCIETY OF MECHANICAL ENGINEERS

UNITED ENGINEERING CENTER / 345 EAST 47TH STREET / NEW YORK, NEW YORK 10017

Statement from By-Laws: The Society shall not be responsible for statements or opinions advanced in papers . . . or printed in its publications (7.1.3)

Authorization to photocopy material for internal or personal use under circumstances not falling within the fair use provisions of the Copyright Act is granted by ASME to libraries and other users registered with the Copyright Clearance Center (CCC) Transactional Reporting Service provided that the base fee of $0.30 per page is paid directly to the CCC, 27 Congress Street, Salem, MA 01970. Requests for special permission or bulk reproduction should be addressed to the ASME Technical Publishing Department.

ISBN No. 0-7918-1049-6

Library of Congress Catalog Number 93-73600

FOREWORD

It is well known that more than sixty percent of all failures in aerospace, transportation and mechanical structures including aerospace vehicles, automobiles, trains, ships and machine components occur due to fatigue. Although considerable progress has been made in the areas of fatigue and fracture of structures and materials, more reliable detection and measurement of flaws and cracks, a better understanding of spectrum load and environmental interaction effects and improved analysis methods for life prediction leads directly to higher performance, reduced weight, reduced cost and longer safe and economic operating life.

This ASME volume on Fatigue and Fracture of Aerospace Structural Materials resulted from papers presented at the symposium on fatigue and fracture held in New Orleans, Louisiana at the 1993 ASME Winter Annual Meeting. The symposium on fatigue and fracture was sponsored by the Structures and Materials Technical Committee of the Aerospace Division of ASME. The papers in this volume present discussions of issues related to fatigue and fracture at high temperatures, damage mechanics and assessment, fracture analysis methods and life prediction. The editors thank the technical and secretarial staff of the Fatigue, Fracture and Reliability Section, Structural Integrity Branch, Flight Dynamics Directorate, Air Force Wright Laboratory for their assistance in many areas of symposium organization and preparation of this ASME volume. Special thanks are also due to the staff of the ASME Technical Publishing Department for their cooperation.

Arvind Nagar
An-Yu Kuo

CONTENTS

AD-Vol. 36, Fatigue and Fracture of Aerospace Structural Materials
ASME 1993

AN ORTHOTROPIC MODEL OF DAMAGE MECHANICS
FOR METAL FATIGUE

C. L. Chow
Department of Mechanical Engineering
University of Michigan, Dearborn
Dearborn, Michigan

L. G. Yu and B. J. Duggan
Department of Mechanical Engineering
University of Hong Kong
Hong Kong

Abstract

This paper presents the development of an orthotropic fatigue damage model to assess life expectancy of engineering components based on theory of Damage Mechanics (DM). This is achieved by extending the application of an anisotropic model of Damage Mechanics recently proposed by Chow and Wang (Chow and Wang, 1987a,b,c, 1988a,b; Wang and Chow, 1989). The effect of plastic damage on fatigue damage is taken into account by establishing the required evolution equations in plastic damage. The failure criterion is based on the postulation that a macro-crack is formed when the overall equivalent damage Z, which includes equivalent fatigue damage Z_f and equivalent plastic damage Z_p, reaches a critical value Z_{cr}. Uniaxial fatigue experiments including single block loading, double block loading and overload on the smooth specimens are carried out to check the validity of the proposed model. The predicted results show good agreement with the experimental data.

1 Introduction

Metal fatigue is a subject of great practical importance as the majority of material failures in engineering components and structures is caused by fatigue. From fundamental physical considerations, metal fatigue results from progressive material degradation accumulated during each cycle commonly known as fatigue damage. During the past 100 years or so, various methods of characterization have been developed to study this problem, some of which are well known, such as the Miner's rule (Miner, 1945) and the Paris' law (Paris, 1957).

In 1958, Kachanov proposed the concept of damage to describe the reduction of load bearing area due to the presence and formation of micro-voids and micro-cracks observed in common engineering materials (Kachanov, 1958). Based on this concept, the Damage Mechanics (DM) has since been extensively developed in recent years (Lemaitre, 1984, 1990; Lee et al, 1985; Chow and Lu, 1989). It not only provides a new approach to study material failures due to plastic deformation, fatigue and creep, but also makes it possible to consider the interactions of multiple damages, which have hitherto been ignored. Recently a generalized model of anisotropic damage has been developed and shown to be capable of solving a wide rang of practical engineering problems including elasto-plastic fracture (Chow and Wang, 1987a,b,c, 1988a,b; Wang and Chow, 1989).

As to the fatigue damage, there are however still some critical engineering problems, as described herein below, which are not well understood and therefore yet to be resolved.

1

1. A limited number of fatigue damage models within the regime of traditional fracture mechanics have been proposed to characterize both fatigue crack initiation and propagation (Kujawske and Ellyin, 1984). But the fact that fracture mechanics deals only with crack propagation makes it clear that neither the energy release rate nor the stress intensity factor has included the consideration of material degradation for crack initiation and crack growth.

2. Most conventional fatigue models (Paris, 1954; Manson, 1965; Coffin, 1954) with or without damage concept, exclude the effect of plastic damage due to monotonic plastic deformation, though an important issue, especially for the cases of multiple block loading and overloading. In fact, material degradation can be caused by either cyclic loading or monotonic loading. The effect of plastic damage should therefore be taken into account.

3. Several fatigue damage models defined in terms of consumed relative fatigue life n_i/N_i such as the Miner's rule (Miner, 1945) were proposed to estimate the residual fatigue life under multiple block loading. But for each block loading, the fatigue life N_i must be determined experimentally, for which the amount of experimental work is considerable, particularly when the effect of nonlinear accumulation is included (Corten and Dolan, 1956). In addition, the accuracy achieved by these models is not considered satisfactory. For the case of overload, the validity of these models become questionable.

4. The fatigue damage model proposed by Chaboche (1988) was established based on the concepts of Damage Mechanics. It was applied to characterize the non-linear accumulation of fatigue damage, the effects of hardening and its interaction with creep damage. But only isotropic damage for structures under uniaxial loading was considered. The idealized model cannot be used to deal with the practical problems including multi-axial load, overload, and propagation rate of a fatigue crack.

In view of the shortcomings in the conventional fatigue models, it becomes desirable for obvious practical reasons to develop a design methodology of fatigue damage capable of taking into account not only the stress amplitude, mean stress, fatigue limit and loading history, but also the progressive material degradation due to the presence of micro-crack and micro-voids. Such a model is developed here by extending the anisotropic damage model due to Chow and Wang (Chow and Wang, 1987a,b,c, 1988a,b; Wang and Chow, 1989), to fatigue crack initiation and propagation under multiple block loading.

2 Basic Concepts

The basic concepts of the anisotropic damage theory, for which the elastic strain energy equivalence hypothesis is shown to be valid, has been described by Chow and Wang (Chow and Wang, 1987a,b,c, 1988a,b; Wang and Chow, 1989), and interested readers may refer to these references for details. However some salient points of the models are briefly illustrated in this section as they are required for the development of the proposed fatigue damage model.

$$\widetilde{S} = M(D):S \tag{1}$$

$$\widetilde{E} = M(D)^{-1}:E \tag{2}$$

$$E = \widetilde{C}^{-1}:S \tag{3}$$

$$W^e(E^e,D) = \frac{1}{2}\widetilde{E}^e:C:\widetilde{E}^e = \frac{1}{2}E^e:\widetilde{C}:E^e \tag{4}$$

$$W^e(S,D) = \frac{1}{2}\widetilde{S}:C^{-1}:\widetilde{S} = \frac{1}{2}S:\widetilde{C}^{-1}:S \tag{5}$$

$$\widetilde{C} = M(D)^{-1}:C:M(D)^{T,-1} \tag{6}$$

2

where $\mathbf{S}$ is the stress tensor, $\widetilde{\mathbf{S}}$ the effective stress tensor, $\mathbf{E}^e$ the elastic strain tensor, $\widetilde{\mathbf{E}}^e$ the effective elastic strain tensor, $W^e(\mathbf{E}^e,\mathbf{D})$ the elastic strain energy, $W^e(\mathbf{S},\mathbf{D})$ the elastic complementary energy, $\mathbf{C}$ the elastic stiffness tensor, and $\mathbf{M}(\mathbf{D})$ the damage effect tensor. In the principal coordinate system of damage, $\mathbf{M}(\mathbf{D})$ is assumed to take the form as follows:

$$\mathbf{M}(\mathbf{D}) = \begin{bmatrix} \dfrac{1}{1-D_1} & & & & & \\ & \dfrac{1}{1-D_2} & & & & \\ & & \dfrac{1}{1-D_3} & & & \\ & & & \dfrac{1}{\sqrt{(1-D_2)(1-D_3)}} & & \\ & & & & \dfrac{1}{\sqrt{(1-D_2)(1-D_3)}} & \\ & & & & & \dfrac{1}{\sqrt{(1-D_2)(1-D_3)}} \end{bmatrix} \tag{7}$$

where $\mathbf{D}$ is the damage tensor. In the principal system, $\mathbf{D}$ is expressed as:

$$\mathbf{D} = \begin{bmatrix} D_1 & 0 & 0 \\ 0 & D_2 & 0 \\ 0 & 0 & D_3 \end{bmatrix} \tag{8}$$

The thermodynamic conjugate force $\mathbf{Y}$ of damage $\mathbf{D}$, also known as elastic strain energy release rate due to damage, is given by

$$\mathbf{Y} = \frac{\partial W^e(\mathbf{E}^e,\mathbf{D})}{\partial \mathbf{D}} = -\mathbf{E}^e:(\mathbf{M}^{-1}:\frac{\partial \mathbf{M}}{\partial \mathbf{D}}:\widetilde{\mathbf{C}})^s:\mathbf{E}^e \tag{9}$$

or

$$\mathbf{Y} = -\frac{\partial W^e(\mathbf{S},\mathbf{D})}{\partial \mathbf{D}} = -\mathbf{S}:(\widetilde{\mathbf{C}}^{-1}\mathbf{M}^{-1}:\frac{\partial \mathbf{M}}{\partial \mathbf{D}})^s:\mathbf{S} \tag{10}$$

in which the superscript s means that only the symmetric part should be taken.
In this paper, $\mathbf{D}$ is defined as the overall damage tensor, including the fatigue damage and plastic damage:

$$d\mathbf{D} = d\mathbf{D}_f + d\mathbf{D}_p = \begin{bmatrix} dD_{f1} & 0 & 0 \\ 0 & dD_{f2} & 0 \\ 0 & 0 & dD_{f3} \end{bmatrix} + \begin{bmatrix} dD_{p1} & 0 & 0 \\ 0 & dD_{p2} & 0 \\ 0 & 0 & dD_{p3} \end{bmatrix} = \begin{bmatrix} dD_1 & 0 & 0 \\ 0 & dD_2 & 0 \\ 0 & 0 & dD_3 \end{bmatrix} \tag{11}$$

where $\mathbf{D}_f$ and $\mathbf{D}_p$ are the fatigue damage tensor and the plastic damage tensor respectively.

3 Fatigue Damage

It is well accepted that the elastic strain energy conventionally defined for monotonic loading is unable to describe the fatigue problem without suitable modifications. The evolution equations of fatigue damage may be derived from the cyclic elastic strain energy $W_c^e(\mathbf{E}^e,\mathbf{D})$ or cyclic elastic complementary energy $W_c^e(\mathbf{S},\mathbf{D})$ defined as:

3

$$W_c^e(\mathbf{E}^e, \mathbf{D}) = \frac{1}{2}\ \widetilde{\mathbf{E}}_c^e : \mathbf{C} : \widetilde{\mathbf{E}}_c^e = \frac{1}{2}\ \mathbf{E}_c^e : \widetilde{\mathbf{C}} : \mathbf{E}_c^e \tag{12}$$

$$W_c^e(\mathbf{S}, \mathbf{D}) = \frac{1}{2}\ \widetilde{\mathbf{S}}_c : \mathbf{C}^{-1} : \widetilde{\mathbf{S}}_c = \frac{1}{2}\ \mathbf{S}_c : \widetilde{\mathbf{C}}^{-1} : \mathbf{S}_c \tag{13}$$

where $\widetilde{\mathbf{S}}_c$ and $\widetilde{\mathbf{E}}_c^e$ are the effective cyclic stress range tensor and the effective cyclic elastic strain range tensor respectively and defined as follows:

$$\widetilde{\mathbf{S}}_c = \mathbf{M(D)} : \mathbf{S}_c = \mathbf{M(D)} : (\mathbf{S} - \mathbf{S}_{me}) = \widetilde{\mathbf{S}} - \widetilde{\mathbf{S}}_{me} \tag{14}$$

$$\widetilde{\mathbf{E}}_c^e = \mathbf{M(D)}^{-1} : \mathbf{E}_c^e = \mathbf{M(D)}^{-1} : (\mathbf{E}^e - \mathbf{E}_{me}^e) = \widetilde{\mathbf{E}}^e - \widetilde{\mathbf{E}}_{me}^e \tag{15}$$

where $\mathbf{S}$ is the stress tensor, $\mathbf{S}_{me}$ the mean stress tensor, $\mathbf{E}^e$ the elastic strain tensor and $\mathbf{E}_{me}^e$ the mean elastic strain tensor. $\mathbf{S}_{me}$ and $\mathbf{E}_{me}^e$ are given by

$$\mathbf{S}_{me} = \frac{1}{2}(\mathbf{S}_{max} + \mathbf{S}_{min}) \tag{16}$$

$$\mathbf{E}_{me}^e = \frac{1}{2}(\mathbf{E}_{max}^e + \mathbf{E}_{min}^e) \tag{17}$$

where $\mathbf{S}_{max}$ and $\mathbf{S}_{min}$ are the maximum stress tensor and the minimum stress tensor respectively; $\mathbf{E}_{max}^e$ and $\mathbf{E}_{min}^e$ are the maximum elastic strain tensor and the minimum elastic strain tensor respectively. For the multi-axial stress state, the terms "maximum" and "minimum" become ambiguous, but they may be regarded as two reverse points of a particular cyclic load range.

The thermodynamic conjugate force $\mathbf{Y}_f$ under fatigue may be obtained from

$$\mathbf{Y}_f = \frac{\partial W_c^e(\mathbf{E}^e, \mathbf{D})}{\partial \mathbf{D}} = -\ \mathbf{E}_c^e : (\mathbf{M}^{-1} : \frac{\partial \mathbf{M}}{\partial \mathbf{D}} : \widetilde{\mathbf{C}})^s : \mathbf{E}_c^e \tag{18}$$

or

$$\mathbf{Y}_f = \frac{\partial W^e(\mathbf{S}, \mathbf{D})}{\partial \mathbf{D}} = -\ \mathbf{S}_c : (\widetilde{\mathbf{C}}^{-1} : \mathbf{M}^{-1} : \frac{\partial \mathbf{M}}{\partial \mathbf{D}})^s : \mathbf{S}_c \tag{19}$$

where $\mathbf{Y}_f$ describes the variation of cyclic elastic strain energy due to fatigue and is chosen to derive the evolution equations of fatigue.

During the nth cycle, the fatigue damage criterion is assumed to be:

$$F_f = Y_{feq}^{1/2} - [C_{f0}(\mathbf{S}_{me}) + C_f(Z_{fn}, Z)] = 0 \tag{20}$$

where $C_{f0}(\mathbf{S}_{me})$ is the initial strengthening threshold, $\mathbf{S}_{me}$ the mean stress tensor, $C_f(Z_{fn}, Z)$ the increment of the threshold, Z_{fn} the equivalent fatigue damage yielded at the nth cycle and Z the equivalent overall damage including the equivalent fatigue damage and the equivalent plastic damage:

$$dZ = dZ_f + dZ_p \tag{21}$$

Y_{feq} is defined as follows:

$$Y_{feq} = \frac{1}{2}\ \mathbf{Y}_f : \mathbf{L}_f : \mathbf{Y}_f \tag{22}$$

where $\mathbf{L}_f$ is the characteristic tensor of fatigue damage which may be expressed as:

4

$$L_f = \begin{bmatrix} 1 & \eta_f & \eta_f & & & \\ \eta_f & 1 & \eta_f & & & \\ \eta_f & \eta_f & 1 & & & \\ & & & 2(1-\eta_f) & & \\ & & & & 2(1-\eta_f) & \\ & & & & & 2(1-\eta_f) \end{bmatrix} \qquad (23)$$

If the fatigue damage criterion of equation (20) is taken to be the irreversible energy dissipation function, the evolution equations of fatigue damage may be accordingly deduced as:

$$\dot{\mathbf{D}}_{fn} = \lambda_f \frac{\partial F_f}{\partial(-\mathbf{Y}_f)} = \frac{-\lambda_f}{2Y_{feq}^{1/2}} \, \mathbf{L}_f \dot{\cdot} \mathbf{Y}_f \qquad (24)$$

$$\dot{Z}_{fn} = \lambda_f \frac{\partial F_f}{\partial(-C_f)} = \lambda_f \qquad (25)$$

where the subscript n indicates the nth cycle. Rewriting equation (24) as

$$\{dD_{fn}\} = \left\{\frac{\partial F_f}{\partial(-Y_f)}\right\} dZ_{fn} \qquad (26)$$

and multiplying it by $\left\{\dfrac{\partial F_f}{\partial D_{fn}}\right\}^T$, we obtain

$$\left\{\frac{\partial F_f}{\partial D_{fn}}\right\}^T \{dD_{fn}\} = \left\{\frac{\partial F_f}{\partial D_{fn}}\right\}^T \left\{\frac{\partial F_f}{\partial(-Y_f)}\right\} dZ_{fn} \qquad (27)$$

From the fatigue damage criterion of equation (20), we have

$$\left\{\frac{\partial F_f}{\partial S_c}\right\}^T \{dS_c\} + \left\{\frac{\partial F_f}{\partial D_{fn}}\right\}^T \{dD_{fn}\} - \frac{\partial C_f}{\partial Z_{fn}} dZ_{fn} = 0 \qquad (28)$$

Substituting equation (27) into equation (28) leads to

$$dZ_{fn} = \frac{\left\{\dfrac{\partial F_f}{\partial S_c}\right\}^T \{dS_c\}}{\dfrac{\partial C_f}{\partial Z_{fn}} - \left\{\dfrac{\partial F_f}{\partial D_{fn}}\right\}^T \left\{\dfrac{\partial F_f}{\partial(-Y_f)}\right\}} \qquad (29)$$

Then substituting equation (29) into equation (26) gives

$$\{dD_{fn}\} = \frac{\left\{\dfrac{\partial F_f}{\partial(-Y_f)}\right\}\left[\left\{\dfrac{\partial F_f}{\partial S_c}\right\}^T \{dS_c\}\right]}{\dfrac{\partial C_f}{\partial Z_{fn}} - \left\{\dfrac{\partial F_f}{\partial D_{fn}}\right\}^T \left\{\dfrac{\partial F_f}{\partial(-Y_f)}\right\}} \qquad (30)$$

At the nth cycle, the accumulated fatigue damage is the sum of the fatigue damage yielded from its previous n-1 cycles, i.e.

5

$$D_f = \sum_{k=1}^{n-1} D_{fk} \tag{31}$$

and

$$Z_f = \sum_{k=1}^{n-1} Z_{fk} \tag{32}$$

where the subscript k indicates the kth cycle.

The instantaneous increment of fatigue damage yielded in one cycle is insignificant as compared with the accumulative fatigue damage D_f over many thousands of cycles and can therefore be ignored. Thus

$$\left\{ \frac{\partial F_f}{\partial D_{fn}} \right\} = \{0\} \tag{33}$$

Integrating equation (29) and equation (30) over one cycle respectively gives

$$Z_{fn} = \frac{dZ_f}{dN} = \int_{S_{cth}}^{S_{cmax}} dZ_{fn} = \int_{S_{cth}}^{S_{cmax}} \frac{\left\{ \frac{\partial F_f}{\partial S_c} \right\}^T \{dS_c\}}{\frac{\partial C_f}{\partial Z_{fn}}} \tag{34}$$

$$\{D_{fn}\} = \frac{\{dD_f\}}{dN} = \int_{S_{cth}}^{S_{cmax}} \{dD_{fn}\} = \int_{S_{cth}}^{S_{cmax}} \frac{\left\{ \frac{\partial F_f}{\partial(-Y_f)} \right\} \left[\left\{ \frac{\partial F_f}{\partial S_c} \right\}^T \{dS_c\} \right]}{\frac{\partial C_f}{\partial Z_{fn}}} \tag{35}$$

where S_{cmax} is the maximum cyclic stress tensor, S_{cth} the threshold of cyclic stress tensor. In the case of uniaxial fatigue, according to the modified Goodman's formula or Gerber's formula (Goodman, 1899; Gerber, 1874), σ_{cth1} is postulated to be:

$$\sigma_{cth1} = \sigma_a \left\{ 1 - \left(\frac{\sigma_{me1}}{\sigma_t} \right)^n \right\} \tag{36}$$

where σ_a is the fatigue limit under symmetric tension-compression loading, σ_{me1} the mean stress and σ_t the tensile strength. When n = 1 equation (36) is reduced to the modified Goodman's relationship, and when n = 2 it is the Gerber's relationship. In this paper, the modified Goodman's relationship is used. For some materials such as aluminum, its endurance limit may be taken as the fatigue limit, σ_{fd}.

In the case of uniaxial fatigue, we have from equations (22), (23) and (19) that

$$Y_{feq}^{1/2} = - \frac{1}{\sqrt{2}} Y_{f1} \tag{37}$$

$$Y_{f1} = - \frac{\sigma_{c1}^2}{E(1-D_1)^3} \tag{38}$$

$$Y_{f2} = Y_{f3} = 0 \tag{39}$$

$$\left\{ \frac{\partial F_f}{\partial S_c} \right\}^T \{dS_c\} = \frac{\sqrt{2} \, \sigma_{c1} d\sigma_{c1}}{E(1-D_1)^3} \tag{40}$$

6

Substituting equation (40) into equation (34), we have

$$\frac{dZ_f}{dN} = \int_{\sigma_{cth1}}^{\sigma_{cmax1}} \frac{\sqrt{2}\,\sigma_{c1}d\sigma_{c1}}{E(1-D_1)^3 \dfrac{\partial C_f}{\partial Z_{fn}}} = \frac{[(\sigma_{max1} - \sigma_{me1})^2 - \sigma_{cth1}^2]}{\sqrt{2}\,E(1-D_1)^3 \dfrac{\partial C_f}{\partial Z_{fn}}} \tag{41}$$

Similarly, for the cyclic fatigue damage, we have

$$\frac{dD_{f1}}{dN} = \int_{\sigma_{cth1}}^{\sigma_{cmax1}} \frac{\sigma_{c1}d\sigma_{c1}}{E(1-D_1)^3 \dfrac{\partial C_f}{\partial Z_{fn}}} = \frac{[(\sigma_{max1} - \sigma_{me1})^2 - \sigma_{cth1}^2]}{2E(1-D_1)^3 \dfrac{\partial C_f}{\partial Z_{fn}}} \tag{42}$$

From equation (41) and equation (42),

$$dZ_f = \sqrt{2}\,dD_{f1} \tag{43}$$

Also it will be shown in equation (66) of a later section that,

$$dZ_p = \sqrt{2}\,dD_{p1} \tag{44}$$

For undamaged material, $D_{f1} = 0$, $Z_f = 0$, $D_{p1} = 0$, $Z_p = 0$, integrating equation (43) and equation (44) gives

$$Z_f = \sqrt{2}\,D_{f1} \tag{45}$$

$$Z_p = \sqrt{2}\,D_{P1} \tag{46}$$

From equation (11), equation (21), equation (45) and equation (46), we have

$$Z = \sqrt{2}\,D_1 \tag{47}$$

During each cycle, if there is no significant plastic damage under a tension-tension load cycle, the increment of overall damage is just the increment of fatigue damage. Introducing equation (47) into equation (41), the explicit solution can be obtained as:

$$\frac{dZ_f}{dN} = \frac{dZ}{dN} = \frac{(\sigma_{max1} - \sigma_{me1})^2 - \sigma_{cth1}^2}{\sqrt{2}\,E\left(1 - \dfrac{Z}{\sqrt{2}}\right)^3 \dfrac{\partial C_f}{\partial Z_{fn}}} \tag{48}$$

which after integration becomes

$$N = \frac{\sqrt{2}\,E}{(\sigma_{max1} - \sigma_{me1})^2 - \sigma_{cth1}^2}\,G_f(Z_{fn}, Z, Z_0) \tag{49}$$

where

$$G_f(Z_{fn}, Z, Z_0) = \int_{Z_0}^{Z}\left(1 - \frac{Z}{\sqrt{2}}\right)^3 \frac{\partial C_f}{\partial Z_{fn}}dZ \tag{50}$$

and Z_0 is the initial value of Z, including the equivalent plastic damage and/or equivalent fatigue damage due to prior monotonic and/or cyclic loading, if any. The

7

fatigue failure criterion governing the threshold condition of a macro-crack initiation is postulated to be

$$Z = Z_{cr} \tag{51}$$

4 Plastic Damage

For the fatigue damage evolution, not only fatigue damage but also plastic damage takes place. It is also well known that fatigue cracks always appear along the persistent slip bands due to localized plastic deformation in smooth specimens (Frost et al., 1974). Therefore, there is a close interaction between plastic and fatigue damages. Accordingly it is necessary to take into account the effect of plastic damage on the evolution of fatigue damage.

The evolution equations of plastic damage may be established in a similar way as those of fatigue damage, by replacing the monotonic elastic strain energy with the cyclic elastic strain energy, from which the plastic damage energy release rate $\mathbf{Y}_p$ may be derived.

$$W^e(\mathbf{E},\mathbf{D}) = \frac{1}{2}\widetilde{\mathbf{E}}^e : \mathbf{C} : \widetilde{\mathbf{E}}^e = \frac{1}{2}\mathbf{E}^e : \widetilde{\mathbf{C}} : \mathbf{E}^e \tag{52}$$

$$W^e(\mathbf{S},\mathbf{D}) = \frac{1}{2}\widetilde{\mathbf{S}} : \mathbf{C}^{-1} : \widetilde{\mathbf{S}} = \frac{1}{2}\mathbf{S} : \widetilde{\mathbf{C}}^{-1} : \mathbf{S} \tag{53}$$

$$\mathbf{Y}_p = - \mathbf{E}^e : (\mathbf{M}^{-1} : \frac{\partial \mathbf{M}}{\partial \mathbf{D}} : \widetilde{\mathbf{C}})^s : \mathbf{E}^e \tag{54}$$

$$\mathbf{Y}_p = - \mathbf{S} : (\widetilde{\mathbf{C}}^{-1} : \mathbf{M}^{-1} : \frac{\partial \mathbf{M}}{\partial \mathbf{D}})^s : \mathbf{S} \tag{55}$$

Thus the plastic damage criterion is:

$$F_p = Y_{peq}^{1/2} - [C_{p0} + C_p(Z)] = 0 \tag{56}$$

where Y_{peq} is defined as,

$$Y_{peq} = \frac{1}{2}\, \mathbf{Y}_p : \mathbf{L}_p : \mathbf{Y}_p \tag{57}$$

where $\mathbf{L}_p$ is the characteristic tensor of plastic damage, which may be expressed as:

$$\mathbf{L}_p = \begin{bmatrix} 1 & \eta_p & \eta_p & & & \\ \eta_p & 1 & \eta_p & & & \\ \eta_p & \eta_p & 1 & & & \\ & & & 2(1-\eta_p) & & \\ & & & & 2(1-\eta_p) & \\ & & & & & 2(1-\eta_p) \end{bmatrix} \tag{58}$$

If the plastic damage criterion of equation (56) is taken to be the irreversible energy dissipation function, the plastic damage evolution equations can be derived similarly,

$$dZ_p = \frac{\left\{\dfrac{\partial F_p}{\partial \mathbf{S}}\right\}^T \{d\mathbf{S}\}}{\dfrac{\partial C_p}{\partial Z} - \left\{\dfrac{\partial F_p}{\partial \mathbf{D}}\right\}^T \left\{\dfrac{\partial F_p}{\partial (-\mathbf{Y}_p)}\right\}} \tag{59}$$

$$\{dD_p\} = \frac{\left\{\frac{\partial F_p}{\partial(-Y_p)}\right\}\left[\left\{\frac{\partial F_p}{\partial S}\right\}^T\{dS\}\right]}{\frac{\partial C_p}{\partial Z} - \left\{\frac{\partial F_p}{\partial D}\right\}^T\left\{\frac{\partial F_p}{\partial(-Y_p)}\right\}} \tag{60}$$

In the case of uniaxial loading,

$$Y_{peq}^{1/2} = -\frac{1}{\sqrt{2}} Y_{p1} \tag{61}$$

$$Y_{p1} = -\frac{\sigma_1^2}{E(1-D_1)^3} \tag{62}$$

$$Y_{p2} = Y_{p3} = 0 \tag{63}$$

$$\left\{\frac{\partial F_p}{\partial S}\right\}^T\{dS\} = \frac{\sqrt{2}\,\sigma_1 d\sigma_1}{E(1-D_1)^3} \tag{64}$$

$$\left\{\frac{\partial F_p}{\partial D}\right\}^T\left\{\frac{\partial F_p}{\partial(-Y_p)}\right\} = \frac{3\sigma_1^2}{2E(1-D_1)^4} \tag{65}$$

From equation (59), equation (60) and equation (61), we have

$$dZ_p = \sqrt{2}\,dD_{p1} \tag{66}$$

In the evolution of plastic damage, the increment of fatigue damage is relatively low and may therefore be neglected. Introducing equation (64), equation (65) and equation (48) into equation (59), we obtain

$$dZ_p = dZ = \frac{2\left(1 - \frac{Z}{\sqrt{2}}\right)\sigma_1 d\sigma_1}{2E\left(1 - \frac{Z}{\sqrt{2}}\right)^4 \frac{\partial C_p}{\partial Z} - 3\sigma_1^2} \tag{67}$$

5 Experimental Investigation

In order to verify the validity of the proposed fatigue damage model described in the preceding sections, a series of experiments was conducted. The material chosen for the fatigue tests was aluminum alloy 2024-T3, whose composition and mechanical properties are as follows: S_i 0.05%, F_e 0.5%, C_u 3.8-4.9%, M_n 0.3-0.9%, M_g 1.2-1.8%, Z_n 0.25%, C_r 0.1%, T_i 0.15%, modulus of elasticity E = 74,343 MPa, yield stress σ_y = 330.0 MPa, tensile strength σ_t = 482.2 MPa, fatigue strength σ_{fd} = 138.2 MPa (5×10^8 cycles).

The experimental data of $C_p(Z_p)$ against Z_p was obtained by Wang (1987), and illustrated in Fig. 1. The fitted Z_p-C_p curve was used in the calculation of plastic damage. The critical value of equivalent overall damage Z_{cr} = 0.1612.

The fatigue specimens of 3.17 thickness shown in Fig. 2 were prepared in such a way that the length was perpendicular to the rolling direction. The severely scratched protective coating made of pure aluminum, whose fatigue strength is about 33% lower than the aluminum alloy, was milled off. After machining, all specimens were polished with fine emery papers from No.240 to No.800.

The test machine is an MTS electro-hydraulic servocontrolled $\pm$10 ton machine. Tests were performed at room temperature. The frequencies of the tests were varied from 5Hz to

9

20Hz.

The experiments were carried out under load-control. All of the specimens were measured before each test with the precision of 0.02 mm. For the two-block fatigue loading and the fatigue tests with overload, the specimens were re-measured after the first load level, or the prior cycles and overloading.

The increment of strengthening threshold of fatigue damage $C_f(Z_{fn}, Z)$ is assumed as follows,

$$C_f(Z_{fn}, Z) = Z_{fn} C_{f1} Z^{C_{f2}} \tag{68}$$

From equation (68), equation (49) and equation (50) we have

$$N = \frac{\sqrt{2} \, E}{(\sigma_{max1} - \sigma_{me1})^2 - (\sigma_{th1} - \sigma_{me1})^2} \int_{Z_0}^{Z} \left(1 - \frac{Z}{\sqrt{2}}\right)^3 C_{f1} Z^{C_{f2}} dZ$$

$$= \frac{\sqrt{2} \, E}{(\sigma_{max1} - \sigma_{me1})^2 - (\sigma_{th1} - \sigma_{me1})^2} [H_f(Z) - H_f(Z_0)] \tag{69}$$

where

$$H_f(Z) = \frac{C_{f1}}{C_{f2}+1} \left\{ Z^{C_{f2}+1}\left(1 - \frac{Z}{\sqrt{2}}\right)^3 + \frac{3}{\sqrt{2}\,(C_{f2}+2)}\left[Z^{C_{f2}+2}\left(1 - \frac{Z}{\sqrt{2}}\right)^2 \right. $$

$$\left. + \frac{2}{\sqrt{2}\,(C_{f2}+3)}\left[Z^{C_{f2}+3}\left(1 - \frac{Z}{\sqrt{2}}\right) + \frac{Z^{C_{f2}+4}}{\sqrt{2}\,(C_{f2}+4)}\right]\right]\right\} \tag{70}$$

Thus for uniaxial fatigue, the solution in an explicit form can be obtained.

Fatigue damage per cycle is difficult to measure precisely and should be deduced from fatigue data under a wide load range of the test system. The least square method was used to determine the parameters determined with equation (69) and the experimental data of eight groups of uniaxial fatigue tests under the single block loading. The fitted parameters are: $C_{f1} = 16,900$ and $C_{f2} = -0.5473$. $C_{f2} < 0$ means that the development of fatigue damage is accelerated with the overall equivalent damage Z.

For these eight groups of test, the range of maximum stress was varied from 252 MPa to 420 MPa with each load increment of either 50 MPa or 20 MPa, while the range of stress amplitude was from 100 MPa to 180 MPa. The Predicted results are shown in Table 1, where σ_{max} is the maximum stress, σ_{min} the minimum stress, N_e the experimental fatigue life, N_p the predicted fatigue life calculated by equation (69), $E_r = (N_p - N_e)/N_e$ the relative error, and N_s the number of specimens used.

Additional uniaxial fatigue tests, including two-block loading and overloading, were carried out to provide an additional verification of the fatigue model. For the two-block loading, two groups of loading were employed, one with High to Low (i.e. 400 to 50 MPa and 300 to 50 MPa) and another with Low to High (i.e. 300 to 50 MPa and 400 to 50 MPa), and other two groups were imposed with the same maximum stress of 350 MPa but with different stress amplitudes of 50 MPa and 150 MPa respectively. The experimental and predicted results are summarized in Table 2 together with those predicted by means of the linear Miner's rule for comparison. In the table, N_{e1} is the cyclic number measured from the first block loading, N_{e2} the experimental fatigue life under the subsequent or second block loading, N_{p2} the predicted fatigue life under the second block loading, E_{rp} the relative error, N_{m2} the predicted fatigue life under the second block loading with the linear Miner's rule and E_{rm} the relative error. The predicted fatigue life was calculated in such a way that the overall equivalent damage Z corresponding to N_{e1} was calculated numerically with equation (69), while the plastic damage, before or after N_{e1}, was calculated numerically with equation (67). Finally, N_{p2} was assessed with equation (69).

For the fatigue with overload, the overload stress imposed was chosen to be 400 MPa. For the first group, the maximum stress was 252 MPa and the stress amplitude was 125 MPa, for which no fatigue pre-loading was applied. The other three groups employed the same maximum stress 300 MPa and stress amplitude 125 MPa, but the cyclic numbers prior to

10

overloading were 0, 60,000 and 120,000 respectively. Though the fatigue overload can be regarded as equivalent to only one-fourth cycle under a high stress above the yield stress, Miner's rule is unable to characterize such a problem. The experimental and predicted results are given in Table 3, where σ_{ov} is the overload stress, N_{pr} the cyclic number prior to overload, N_e the experimental fatigue life after overload, N_p the predicted fatigue life after overload, E_r the relative error. The method of prediction based on the proposed fatigue damage model was similar to that of two-block fatigue.

The number of specimens for each group employed varied from 4 to 9, which was equal to or larger than the minimum number of specimens recommended for fatigue studies (Gao, 1986). The confidence level achieved was $\gamma = 95\%$ and the relative error limit, $\delta = 5\%$.

6 Conclusions and Discussion

An anisotropic fatigue damage model is presented, including the development of the evolution equations of plastic damage and fatigue damage. The characterization of these damages requires the experimental determination of two parameters, C_{f1} and C_{f2} for the fatigue damage. The predicted results show good agreement with the experimental data. Based on the model, the damage accumulation rate depends on the amount of accumulated damage, capable of taking into account the effect of the loading history, including monotonic loading and cyclic loading.

From the viewpoint of Damage Mechanics, there is no need to employ different criteria for crack initiation and crack propagation, both based on the postulation that the threshold condition of the failure of an element with certain characteristic dimensions is satisfied when $Z = Z_{cr}$. With the aid of Finite Element Method, which should include the effects of damage, the proposed model can also be used to calculate fatigue crack initiation and propagation under multi-axial stress.

The effects of monotonic plastic deformation on fatigue damage are complex, and are governed by two aspects, namely plastic damage and plastic hardening. The former reduces fatigue life but the latter may enhance fatigue life (Frost et al., 1974). However, the theory of Damage Mechanics provides a method capable of assessing design life of engineering components under multiple loading environments, including plastic deformation, fatigue and creep.

References

Chaboche, J.L., and Lesne, P.M., 1988, "A Non-linear Continuous Fatigue Damage Model", *Fatigue and Fracture of Engineering Materials and Structures*, Vol. 1, pp. 1-17.

Chow, C.L., and Lu, T.J., 1989, "On Evolution Laws of Anisotropic Damage", *Engineering Fracture Mechanics*, Vol. 34, pp. 679-701.

Chow, C.L., and Wang, J., 1987a, "An Anisotropic Theory of Elasticity for Continuum Damage Mechanics", *International Journal of Fracture*, Vol. 33, pp. 3-16.

Chow, C.L., and Wang, J., 1987b, "An anisotropic Theory of Continuum Damage Mechanics for Ductile Fracture', *Engineering Fracture Mechanics*, Vol. 27, pp. 547-558.

Chow, C.L., and Wang, J., 1987c, "An Anisotropic Continuum Damage Theory and Its Application to Ductile Crack Initiation", *Damage Mechanics in Composites*, ASME AD-Vol. 12, pp. 1-10.

Chow, C.L., and Wang, J., 1988a, "Ductile Fracture Characterization with an Anisotropic Continuum Damage Theory', *Engineering Fracture Mechanics*, Vol. 30, pp. 547-563.

Chow, C.L., and Wang, J., 1988b, "A Finite Element Analysis of Continuum Damage Mechanics for Ductile Fracture", *International Journal of Fracture* Vol. 38, pp. 83-102.

Coffin, J.F., Jr, 1954, *Transactions* ASME, Vol. 76, p. 931.

Corten, H.T., and Dolan, T.J., 1956, "Cumulative Fatigue Damage", *Proceedings, International Conference on Fatigue of Metals*.

Frost, N.E., Marsh, K.J., and Pook, L.P., 1974, Metal Fatigue, Clarendon press, Oxford.

Gao, Z.T., 1986, "Applied Statistics of Fatigue", (in Chinese), National Defense Industry Press.

Gerber, W.Z., 1874, Bayer. Achit. Ing. Ver. 6, p. 101.

Goodman, J., 1899, Mechanics Applied to Engineering, Longman, Green, and Company, London.

Kachanov, L.M., 1958, *Izvestiya Akademii Nauk . SSSR. Otd. Tekhn. Nauk.*, Vol. 8, pp. 26-31.

Kujawske, D., and Ellyin, F., 1984, "A Cumulative Damage Theory for Fatigue Crack

Initiation and Propagation", *International Journal of Fatigue*, Vol 2, pp. 83-88.

Lee, H., Peng, K., and Wang, J., 1985, "An Anisotropic Damage Criterion for Deformation Instability and Its Application to Forming Limit Analysis of Metal Plate", *Engineering Fracture Mechanics*, Vol. 21, pp. 1031-1054.

Lemaitre, J., 1984, "How to Use Damage Mechanics", *Nuclear Engineering Design*, Vol. 80, pp. 233-245.

Lemaitre, J., 1990, "Micro-mechanics of Crack Initiation", *International Journal of Fracture*, Vol. 42, pp. 87-89.

Manson, S.S, 1965, *Experimental Mechanics*, Vol 5, p. 193.

Miner, M.A., 1945, *Journal of Applied Mechanics*, Vol. 12, A159 .

Paris, P.C., 1957, The Boeing Company, Document No. 17867, Addendum N.

Wang, J., 1987, "Development of An Anisotropic Damage Mechanics Model in Ductile Fracture", Ph.D. Thesis, University of Hong Kong.

Wang, J., and Chow. C.L., 1989, "Mixed Mode Ductile Fracture Studies with Non-proportional Loading Based on Continuum Damage Mechanics", *Journal of Engineering Materials and Technology,* Vol. 111, pp. 204-209.

Table 1. Fatigue under Single-block Loading

σ_{max}	σ_{min}	N_e	N_p	E_r	N_s
420.0	60.0	26,700	24,300	-0.054	5
420.0	170.0	62,100	52,600	-0.153	4
400.0	50.0	33,200	30,700	-0.075	4
400.0	150.0	80,500	63,800	-0.207	4
350.0	50.0	61,100	70,100	0.147	4
350.0	150.0	188,100	200,900	0.068	4
300.0	50.0	197,600	196,500	-0.006	4
252.0	2.0	266,200	292,300	0.098	6

Table 2. Fatigue under Double-block Loading

Level 1			Level 2			N_{p2}	E_{rp}	N_{m2}	E_{rm}	N_s
σ_{max}	σ_{min}	N_{e1}	σ_{max}	σ_{min}	N_{e2}					
400.0	50.0	15,000	300.0	50.0	49,800	50,100	0.006	108,300	1.175	4
300.0	50.0	100,000	400.0	50.0	22,400	19,900	-.110	16,400	-0.268	4
350.0	150.0	94,000	350.0	50.0	35,600	37,300	0.048	30,600	-0.140	4
350.0	50.0	30,600	350.0	150.0	96,500	113,300	0.174	94,000	-0.026	9

Table 3. Fatigue with Overload

σ_{max}	σ_{min}	σ_{ov}	N_{pr}	N_e	N_p	E_r	N_s
252.0	2.0	400.0	0	136,800	146,800	0.073	4
300.0	50.0	400.0	0	108,700	98,100	-0.097	4
300.0	50.0	400.0	60,000	84,900	85,600	0.007	4
300.0	50.0	400.0	120,000	54,200	50,500	-0.069	7

(all stresses in MPa)

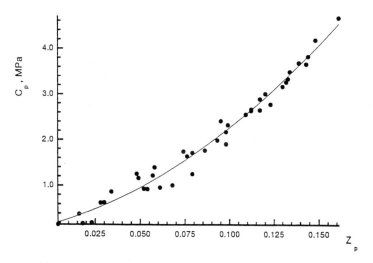

Fig. 1 Damage strengthening curve of aluminium alloy 2024-T3

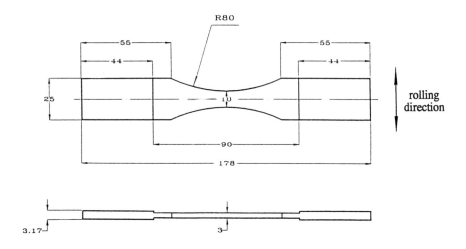

Fig. 2 Smooth fatigue specimen

AD-Vol. 36, Fatigue and Fracture of Aerospace Structural Materials
ASME 1993

HYDROGEN ENHANCED LOCALIZED PLASTICITY:
A MECHANISM FOR HYDROGEN RELATED FRACTURE

P. Sofronis
Department of Theoretical and Applied Mechanics
H. K. Birnbaum
Department of Materials Science and Engineering

University of Illinois, Urbana–Champaign
Urbana, Illinois

ABSTRACT

The mechanisms of hydrogen related fracture are briefly reviewed and a few evaluative statements are made about the stress induced hydride formation, decohesion, and hydrogen enhanced localized plasticity mechanisms. A more complete discussion of the failure mechanism based on hydrogen enhanced dislocation mobility is presented and these observations are related to measurements of the macroscopic flow stress. The effects of hydrogen induced slip localization on the measured flow stress is discussed. A theory of hydrogen shielding of the interaction of dislocations with elastic stress centers is outlined. It is shown that this shielding effect can account for the observed hydrogen enhanced dislocation mobility.

REVIEW OF PERTINENT OBSERVATIONS

Despite extensive study, the mechanism(s) of hydrogen embrittlement has remained unclear. Several candidate mechanisms have evolved, each of which is supported by sets of experimental observations and strong personal views. One reasonable certain aspect of this controversy is that there are several viable mechanisms of hydrogen related failure and that the search for a single mechanism to explain all observations is doomed to failure. Of the many suggestions, three mechanisms appear to be viable; stress induced hydride formation and cleavage [1-4], hydrogen enhanced localized plasticity [5-9], and hydrogen induced decohesion [10,11]. The first of these has been definitively established to be operative in systems in which hydrides are either stable, or can be stabilized by the application of a stress field, e.g. Group Vb metals [3, 12-14], Ti [4, 15], and, Zr [16]. This "second phase" mechanism is supported by microscopic observations [11, 17] and thermodynamic calculations [18]. In these hydride forming systems it has been shown [4] that under conditions in which the hydride cannot form, hydrogen "embrittlement" will occur by the second mechanism named above, hydrogen enhanced localized plasticity.

The hydrogen enhanced localized plasticity mechanism is based on observations that in a range of temperatures and strain rates, the presence of hydrogen in solid solution decreases the barriers to dislocation motion, thereby increasing the amount of deformation that occurs in a localized region adjacent to the fracture surface [19-25]. The fracture process is a highly localized plastic failure process rather than an embrittlement. This counterintuitive process says that the macroscopic ductility is limited by the onset of extensive localized plasticity and is supported by microscopic observations. The third viable mechanism is the hydrogen related decohesion mechanism in which the atomic bonding at the crack tip is weakened by the presence of hydrogen in solid solution [10,11]. This mechanism is supported primarily by the observations that in some non-hydride forming systems, hydrogen embrittlement appears to occur in the absence of significant local deformation, by theoretical calculations of the effect of hydrogen on the atomic potentials [26] and by a thermodynamic argument [27, 28]. Direct evidence for this mechanism has not been obtained and measurements which have been made on the effects of hydrogen on small strain aspects of the lattice potential suggest no softening of the lattice potential [7].

In the present paper we will review the observations supporting the hydrogen enhanced localized plasticity (HELP) mechanism of hydrogen embrittlement and then discuss the mechanism by which hydrogen enhances the mobility of dislocations. High resolution fractography of hydrogen embrittled metals, such as Ni and Fe, show extensive plastic deformation localized along the fracture surfaces [25, 29]. Particularly revealing results have been obtained with the technique of *in situ* environmental cell deformation and fracture. These methods allow observation of the fracture process in real time, at high spatial resolution, and in vacuum or in H_2 atmospheres. These studies have been carried out for bcc, fcc, and hcp metals having various solute contents and for solid solutions, precipitation strengthened alloys, and intermetallics [19-24, 30]. In systems which exhibit hydrogen embrittlement, the nature of hydrogen effects, while differing in details, were the same in fundamental character.

The most dramatic observation was that hydrogen increased the dislocation mobility under conditions of constant stress. In stressed specimens, dislocation mobility could be increased by the addition of H_2 to the environmental cell and decreased by removal of H_2 and restoration of vacuum. This behavior was observed for edge, screw, and mixed dislocations and for isolated dislocations as well as dislocation tangles. In bcc metals the enhanced dislocation velocities were observed on {112} and {110} slip planes and the enhancement was least for extremely high purity Fe and increased as interstitial solutes (C) were added to the solid solution. Hydrogen enhanced dislocation velocity was observed for dislocations completely contained within the specimens as for Frank-Read sources and for dislocations which terminated at the surfaces. Hydrogen enhanced operation of dislocation sources were observed within the crystals and at grain boundaries.

Specimens which contained stress concentrations, such as notches, failed by ductile plastic processes at the front of the notch when stressed in Mode 1 in vacuum. When stressed under gaseous H_2, similar fracture was observed, with the exception that the extent of plasticity was more confined to regions adjacent to the fracture surface in the case of fracture in H_2. In relatively pure specimens (Ni. Fe, etc.) the fracture tended to be along slip planes and the deformation accompanying the fracture in H_2 was within 1 micrometer of the active slip plane. In alloys such as stainless steels [23], the fracture in H_2 was much less crystallographic and tended to be along the plane of maximum normal stress.

16

In all of the systems, the cracks which had propagated in vacuum and then stopped under a constant external load could be started and continued to propagate without any increase in external load when H_2 was added to the environmental cell. This process occurred by increasing the dislocation activity at the crack tip when the specimen was exposed to H_2 gas. Since the fracture process in H_2 was by localized ductile failure e.g. by the formation of very shallow, localized microvoids [31], the increase in the dislocation velocity at the crack tip is the root cause of the hydrogen embrittlement. The effect of hydrogen is greatest at the crack tip where either hydrogen entry is facilitated by slip processes or the local hydrogen concentration is increased by the effect of stress on the chemical potential of the solute H. Hydrogen locally softens the material in front of the crack allowing ductile fracture to occur there, prior to general yielding away from the crack tip. In the cases where the concentration of hydrogen is greatest in the vicinity of grain boundaries [32-34], it is in these regions that deformation occurs at the lowest stresses and hence "intergranular" fracture is observed. In cases where high resolution studies of these "intergranular" fracture have been carried out [4, 35] it is clear that the fracture in fact occurs by plastic processes in the vicinity of the boundaries rather than along the boundaries themselves.

HYDROGEN EFFECTS ON MACROSCOPIC DEFORMATION

Despite the many observations of hydrogen enhanced dislocation mobility in TEM specimens, relatively few observations of softening due to hydrogen in solid solution have been reported. In most cases, the introduction of hydrogen has been reported to result in increased flow stress and only in a small number of cases has softening been observed [36-39]. Significant decreases of the flow stress have been reported in high purity iron which was cathodically charged at very low current densities and during hydrogen charging under conditions in which the hydrogen was introduced without damage to the specimen. Two significant differences appear to divide the experiments which exhibit softening due to hydrogen from those which do not. Hydrogen decreases have been observed at very low strain rates and when hydrogen was introduced under conditions which did not cause any structural damage to the specimens. The latter condition is particularly significant, as most cathodic charging conditions used to introduce high supersaturations of hydrogen introduce high concentration gradients and correspondingly high stresses near the surface. The consequent deformation in the near surface region gives rise to hardening which may mask the softening caused by hydrogen. In the case of the cathodic charging conditions utilized by Kimura et al [36-38], no surface deformation was observed.

The effects of hydrogen on the thermal activation of dislocations over barriers has been studied using the techniques of load relaxation, temperature and strain rate changes in high purity Ni and Ni-C alloys [40]. The critical parameters which control the thermal activation of dislocations over barriers are the activation volume, A^*, and the activation enthalpy, ΔH^*. These parameters were determined using stress relaxation and temperature cycling techniques on Ni and Ni-C alloys [40]. As expected from the strong strengthening effects of C in solid solution in Ni, solute C increases A^* over a wide stress range and increases ΔH^* over the same stress range. This is consistent with C solutes or small C clusters causing strong barriers to the dislocation motion. In contrast, solute H decreases A^* and decreases ΔH^* consistent with a hydrogen related decrease in the strength of the dislocation obstacles. Adding hydrogen solutes to a Ni-C alloy has the effect of approximately removing the increased v^* and H^* caused by C, i.e. removing the enhanced dislocation obstacles caused by C. These results can be used to calculate the effects of solute H on

17

dislocation velocities with the results shown in Table 1 where σ_a is the applied stress, $\delta\Delta H^*$ is the hydrogen induced change in activation enthalpy, $\delta A^*/b^2$ is the hydrogen induced change in activation area, and v_D^H / v_D is the dislocation velocity in the presence of H relative to that in the absence of H as calculated from the measured dislocation parameters.

Table 1. EFFECT OF HYDROGEN ON THE ACTIVATION ENTHALPY, ACTIVATION AREA AND DISLOCATION VELOCITY

	σ_a (MPa)	$\delta\Delta H^*$ (eV)	$\delta A^*/b^2$	v_D^H / v_D
Pure Ni	50	-0.32	-500	3×10^5
Pure Ni	100	-0.11	-200	81
Pure Ni	150	-0.05	-100	7
Ni-C	50	-0.49	-800	3×10^8
Ni-C	100	-0.13	-300	181
Ni-C	150	-0.06	-150	11

The values of v_D^H / v_D indicate that significant enhancement of the dislocation velocities are expected in Ni-H and Ni-C-H alloys, particularly at the lower applied stresses. This is in accord with the *in situ* TEM studies in which the stresses, while not directly measured, are expected to be in the range of 100-150 MPa. These results indicate that the H related softening is greatest at the lower applied stresses where the dislocation velocities are the smallest.

SLIP LOCALIZATION

Measurements of H induced changes in the dislocation activation parameters using stress relaxation and thermal cycling measurements, were carried out on bulk specimens. While they have not been carried out on as wide a range of materials as the TEM studies, they indicate that the *in situ* TEM studies are not artifacts of the use of thin specimens. There does remain the question of why many studies, which measure macroscopic stress-strain curves, exhibit increases in the measured flow stresses due to the introduction of hydrogen, whereas the TEM specimens generally show either no effect (in materials which do not absorb significant amounts of hydrogen) or an enhanced dislocation velocity due to the introduction of hydrogen. One factor, discussed briefly above, is the introduction of near surface damage due to the H charging of macroscopic specimens, which would cause hardening in a flow stress determination. This damage, which results from the severe hydrogen concentration gradients (and accompanying stress gradients) which can often accompany cathodic charging, will depend on the details of the charging conditions, the surface condition, and the purity of the material (through the propensity for impurities to trap hydrogen, decrease the diffusivity and increase the near surface concentration gradients).

Another factor is the localization of slip on a limited number of slip planes due to hydrogen charging as observed in stainless steels [41], Ni [42], and high purity Al [39]. In the stainless steel-H alloys the strain localization was accompanied by significant increases in the flow stress, while in the Al-H alloys, significant decreases in the flow stresses was observed. Shear localization occurs if the flow stress in the region of localization is reduced <u>relative</u> to the flow stress of the non-deforming volume.

This can occur in several ways and a number of these cases are considered below.

Case I. In systems where H reduces the barriers to dislocation motion, an inhomogeneous distribution of H can cause shear localization since the flow stress is lower where the H concentration is greatest. This may occur if H entry is facilitated at slip line intersections with the surface.

Case II. Shear localization can occur when the specimen is hardened by the introduction of hydrogen (due to the formation of hydrides or H clusters), if the initial deformation locally reduces the stress for continued slip. This is believed to be the case for the Al-H system where the hardening of the crystal is believed to occur by the formation of small hydrogen clusters [43]. Removal of these clusters during the initial passage of dislocations leads to slip localization as in other precipitation hardened systems.

Increases in dislocation mobility due to the presence of solute H will enhance this slip localization in both cases. The macroscopic flow stress can be increased or decreased, depending on the magnitude of the local softening due to the removal of dislocation barriers. In the case of stainless steel, severe slip localization is accompanied by significant increases in the macroscopic flow stress [41] while in the case of Al the macroscopic flow stress is reduced [39].

In general, measurements of the macroscopic stress strain behavior which show increases in the flow stress due to the introduction of H may result from shear localization even in the presence of reduced barriers to dislocation motion.

ELASTIC SHIELDING OF STRESS CENTERS

Since the hydrogen related softening was observed in fcc, bcc, and hcp systems, in pure metals, solid solutions, and in precipitation hardened systems, it appears to be a very general effect. Hence the cause for this softening should be sought in the elastic interactions between dislocation and between dislocations and other stress centers, such as solute atoms and precipitates. It is this elastic interaction which provides the common strengthening mechanism in all these systems. Hydrogen atmospheres form in the dilatational fields of the elastic singularities; in the tension field of the edge dislocations, in the volume surrounding solutes (in particular interstitial solutes), and in the vicinity of precipitates. The primary cause of the interaction which gives rise to the H atmospheres is the first order dilatational interaction energy, W_{int} which can be expressed as:

$$W_{int} = -\frac{1}{3}\sigma_{kk}^{a}\Delta v \qquad (1)$$

where σ_{kk}^{a} is the stress field of a defect and Δv is the unconstrained volume dilatation of the H solute. For the interaction of H with an edge dislocation this may be written:

$$W_{int} = -\frac{\mu b(1+\upsilon)\sin\phi}{3\pi(1-\upsilon)r}\Delta v \qquad (2)$$

where μ and υ are the shear modulus and Poisson's ratio respectively, b is the magnitude of the Burgers vector, and ϕ and r are the polar coordinates as measured from the dislocation core and slip plane. As a result of this elastic interaction, the concentration of H in the atmosphere is described by [44]:

$$\frac{c_H}{1-c_H} = \frac{c_o}{1-c_o} \exp\left(-\frac{W_{int}}{kT}\right) \qquad (3)$$

where c_o is the H concentration far from the defect in a zero stress field, and c_H is the H concentration in the atmosphere. In addition to this first order elastic interaction, there is a second order elastic interaction energy which arises from the change in the elastic moduli caused by the presence of H in solid solution. For the case of the Nb-H system:

$$\mu = \mu_o \frac{1+0.34c_H}{1-0.0177c_H} \;,\; \upsilon = \upsilon_o - 0.025c_H \;,\; E = E_o(1+0.34c_H). \qquad (4)$$

The second order elastic interaction between H and the dislocation may be written as:

$$W'_{int} = \frac{1}{2}\varepsilon_{ij}\varepsilon_{kl}\left(c'_{ijkl} - c_{ijkl}\right) \qquad (5)$$

where the primed elastic constants refer to those in the presence of H in solid solution.

The effects of H on the elastic interactions between defects have been calculated using a linear elasticity and finite element techniques [45] and some of the results are summarized below. Shielding of the elastic interactions results from the fact that the hydrogen atmospheres around both the dislocation and the interacting stress centers respond to the total stress at each point. The total elastic force between the defects must include the interaction forces between the dislocation and the elastic center, between the H atmospheres, and between the H atmospheres and the dislocation and elastic stress center.

In the absence of hydrogen, the interaction force per unit length between two parallel dislocations is given by:

$$F = \sigma_1 b_2 \times \xi \qquad (6)$$

where σ_1 is the stress tensor of dislocation 1, b_i with i=1,2, are the Burgers vectors, and ξ is the unit vector along the dislocation lines. In component form, Eqn. 6 is written as $F_i = \varepsilon_{ijk}(\sigma_1)_{jm}(b_2)_m \xi_k$ where ε_{ijk} is the alternating symbol. In the case to be discussed, parallel edge dislocations, Eqn. 6 results in a repulsive force if the dislocations are of the same sign and on the same slip system. In this case, the force on dislocation 2 due to dislocation 1 is $\tau_D b_2$ where τ_D is the stress of dislocation 1 resolved along the slip plane and Burgers vector.

The hydrogen effect on the interaction between the dislocations 1 and 2 (Fig. 1) is assessed by calculating the hydrogen induced change in the shear stress τ_D due to interactions between the hydrogen atmospheres surrounding the two dislocations. These hydrogen atmosphere are modeled by a continuous distribution of dilatation lines parallel to the dislocation lines. Each dilatation line is viewed as a stress source which affects the shear stress τ_D. This type of model for the hydrogen effect is consistent with the plane strain assumption for the dislocation strain field when the hydrogen concentration does not vary in the direction of the dislocation lines. In that case the in-plane concentration of the dilatation lines, n, is directly related to the hydrogen concentration per unit volume, C through n=Ch where h is the distance between two successive hydrogen atoms along the dilatation line and the concentration n denotes the number of hydrogen atoms per unit area in the plane normal to the dilatation line.

The net shear stress, τ_H, induced by the hydrogen atmospheres is found by integration of the stress contributions of each of the H dilatation lines over the entire area S occupied by the atmosphere. In polar coordinates, the shear stress on dislocation 2 due to the H atmospheres is given by:

$$\tau_H = -\frac{\mu}{2\pi(1-v)}\frac{V_H}{N_A}\int_0^{2\pi}\int_{r_2}^R C(r,\phi)\frac{\sin 2\phi}{r}\,dr\,d\phi \qquad (7)$$

where r_2 is the inner cutoff radius of dislocation 2 and R is the outer cutoff radius of the atmosphere centered at dislocation 2. The core of dislocation 1 with cutoff radius r_1 is also excluded from the integration. The elastic solution shows that the stress field of a hydrogen dilatation line is purely deviatoric. Consequently, the interaction energy between the hydrogen dilatation lines is zero and introduction of a dilatation line into the lattice is energetically independent of the presence of the neighboring lines. Therefore the hydrogen concentration $C(r,\phi)$ at any position (r,ϕ) is determined solely by the corresponding stress due to dislocations 1 and 2. Superposition of the singular linear elastic stress fields of the two dislocations yields the in-plane hydrostatic stress as:

$$\frac{\sigma_{11}^a + \sigma_{22}^a}{2} = -\frac{\mu}{2\pi(1-v)}\left(b_2\frac{\sin\phi}{r} + b_1\frac{r\sin\phi - l\sin\omega}{r^2 + l^2 - 2rl\cos(\phi-\omega)}\right) \qquad (8)$$

where l and ω are the polar coordinates of the position of dislocation 1 (see Fig. 1). Then concentration $C(r,\phi)$ at every point in the solid is calculated using the total stress field, Eqn. 8.

Equation 7 indicates that the stress field of a hydrogen dilatation line decays as $1/r^2$ with distance r. It is expected that the magnitude of the shear stress due to H, τ_H, will depend mainly on the dilatation lines close to the core of dislocation 2. The corresponding shear stress, τ_D, at the core of dislocation 2 due to dislocation 1 is given by

$$\tau_D = -\frac{\mu b_1}{2\pi(1-v)}\frac{\cos\omega\cos 2\omega}{l} \qquad (9)$$

and the net shear stress exerted on dislocation 2 is equal to $\tau_D + \tau_H$.

Calculations of hydrogen effects on the dislocation interactions were carried out using the parameters appropriate to Nb, as there is a full set of data available for the Nb-H system. The hydrogen atmosphere in equilibrium with the stress field of a single dislocation in an infinite medium is shown in the form of normalized iso-concentration lines, C/C_0, in Fig. 2 at a nominal concentration $C_0=0.1$ hydrogen atoms per solvent atom and a temperature of 300K. The atmosphere is symmetric with respect to the dislocation plane because of the corresponding symmetry in the hydrostatic stress field of the dislocation. Under the same temperature and nominal hydrogen concentration, Fig. 3 shows the hydrogen atmosphere for dislocations 1 and 2 on the same slip system and at respective relative positions of 10, 8 and 6 Burgers vectors apart. The portion of the hydrogen atmosphere round each dislocation is non-symmetric when compared with the atmosphere of the single dislocation shown in Fig. 2 as a direct result of the linear superposition in the stress field of the two dislocations.

In Fig. 4 the hydrogen induced shear stress, τ_H, at the core on the slip plane of dislocation 2 is plotted against distance l between the dislocations at H/M=0.1, 0.01 and 0.001 and at a temperature of 300K. The angle ω is equal to 180° (see Fig. 1). The shear stress is normalized by the shear modulus μ and the distance by the Burgers vector magnitude b. In the same figures the normalized shear stress, τ_D/μ, due to dislocation 1 in the absence of hydrogen and the total normalized shear stress, $(\tau_D +\tau_H)/\mu$, are plotted as well.

The shear stress on dislocation 2, due to hydrogen, is negative and its absolute value increases as the nominal H concentration, c_o, becomes larger, consistent with concentration dependence of the H line density. In both the interactions between like signed dislocations and unlike signed dislocations which were also calculated [45] the effect of hydrogen is to decrease the force exerted on dislocation 2 by dislocation 1.

The effects of hydrogen shielding on the interaction of edge and screw dislocation with solutes having lower symmetry that the lattice, e.g., interstitials in bcc metals [46], have been calculated using a finite element method in a self consistent procedure which minimizes the energy of the system while calculating the H distribution around the dislocation and the solute atom. In the absence of H, the method gives good agreement with the analytic calculations of Cochardt et al. [46]. In the presence of H, the complete elastic solution is satisfied with variable moduli which depend on the local H concentration. The results of these calculations for the interactions of an edge dislocation lying along the [211], with its Burgers vector in the [1$\bar{1}$1], with C interstitials having their tetragonal axes along the three <100> are shown in Figs. 5-7. The analytic result is in good agreement with the previous calculations [46] and the effect of H in solid solution is to markedly decrease the interaction energy for all three C orientations. For all three C orientations, the presence of H causes a decrease in the edge dislocation - C interaction energy of about 0.5 eV. The decrease in the interaction energy with H is consistent with the measured decrease in the activation enthalpy, ΔH^*, for thermally activated deformation [40]. Similarly, the presence of H gives rise to a decrease in the width of the interaction potential well which is consistent with the measured decrease in the activation area, A^*, in the presence of solute H as determined by the thermally activated flow measurements [40].

DISCUSSION

The consequence of hydrogen shielding of the stress fields of elastic singularities are numerous and will be explored in a subsequent publications. At the present time, we wish simply to point out some of the situations which will be of greatest interest for the issues of hydrogen related fracture. In all cases, the shielding effects are manifested at low strain rates and at temperatures where the H solutes retain a high mobility. In the absence of hydrogen, it is well known that two parallel edge dislocations on the same slip system and with Burgers vectors of the same sign repel each other. Both the analytical and finite element calculation of H shielding indicate that the shielding greatly reduces the repulsive force acting between the dislocations on the glide plane. This reduction is associated with the volumetric strain produced by the introduction of hydrogen into the lattice, (volumetric effect), and the hydrogen induced changes in the constitutive moduli, (modulus effect). The hydrogen related decrease of the repulsive interaction depends strongly on the nominal hydrogen concentration, as indicated by Fig 4. As expected from the fact that the H shielding effect results from the stress fields of point defects, it affects the interaction between dislocations at short range, in agreement with the $1/r^2$

dependence of the stress field of the interstitial hydrogen. At sufficiently high H concentrations, the repulsive stress field at short range can become attractive due to the H stress fields, possibly leading to coalescence of the leading dislocations at the head of a pileup and the formation of a crack in the manner suggested by Stroh [47]. A similar reduction of the attractive force between opposite signed dislocations leads to a weakening of the slip barriers provided by dislocation dipoles. Similarly, the forces acting on dislocations which cause glide polygonization are decreased by the presence of solute H atmospheres, leading to a more random dislocation distribution during deformation and to a lowering of the internal stress fields.

The most significant effect seems to be a decrease in the strength of the interactions between dislocations and point defects and between dislocations and internal stress fields provided by other dislocations due to the H shielding, as shown in Figs. 4-7. Both of these leads to an increased dislocation velocity under stress as the slip barriers can be more easily overcome by thermal activation. These effects are shown schematically in Fig. 8 for various temperatures and strain rates. The effects of solute H are expected to be small at elevated temperatures where no significant H atmospheres can form at the dislocations or at the barriers to slip. "High temperature" behavior is expected when $kT > B$, the "binding enthalpy" of H to the dislocation. Since the effects of H shielding are manifest only if the H atmospheres can move with the dislocations, little effects due to shielding are expected at "low temperatures". The definition of "low temperature" depends on the dislocation velocity and hence the imposed strain rate and mobile dislocation density. In the "low temperature" range a small yield point may be expected as the pinning of the dislocations by the immobile H atmospheres limits the initial mobile dislocation density and allows the dislocations to exhibit "breakaway effects". Both decreases and increases in dislocation velocities due to H atmospheres can be manifested at "intermediate temperatures" where the H can move with the dislocations. At high strain rates, where the H atmospheres move with, but lag behind the dislocations, decreases in dislocation mobility and macroscopic hardening may be expected as the atmospheres provide a drag force. At some what higher temperatures, serrated yielding (Portevin - LeChatalier effect) can be expected and has been observed [48]. Increases in dislocation mobility due to H shielding are expected at each strain rate above the temperature of serrated yielding, as it is in this temperature and strain rate range that the H atmospheres can move with the dislocations and rearrange themselves to provide maximum shielding of the elastic interactions. It is this intermediate temperature range, in which H enhanced dislocation mobility is expected, that corresponds to the temperature range in which H embrittlement due to solute H is generally observed.

While some macroscopic deformation observations can be considered as supportive of the direct observations of H enhanced dislocation mobility, in many similar measurements, flow stress hardening rather than softening is observed. Several reasons may account for this dichotomy.

a. Many of the macroscopic deformation measurements which exhibit hardening on the introduction of H have been carried out at relatively high strain rates and at temperatures where the lagging of the H atmospheres behind the moving dislocations cause increases in the flow stress.

b. As discussed above, H induced slip localization can cause increases in the measured flow stress in experiments under imposed strain rates; even under conditions where the dislocation mobility is increased by the solute H.

c. Damage to the material during the introduction of H by high fugacity charging can result from very large H concentration gradients which cause stress gradients during charging. While small amounts of dislocation damage can cause softening due to introduction of dislocation sources, in general the large near surface deformations cause hardening of the material. This is particularly true in cases such as Ni where surface hydrides can form during cathodic charging.

CONCLUSIONS

In view of the complexity of H effects in the deformation and fracture of solids, no firm conclusions about the mechanism(s) of hydrogen related fracture can yet be reached. The evidence for fracture due to stress induced hydrides under conditions where hydrides are stable under stress is incontrovertible. Less firm conclusions can be reached for systems which do not exhibit hydride formation.

In non-hydride forming systems, the evidence for failure by Hydrogen Enhanced Local Plasticity is strong, based on microscopic observations of crack tip behavior and fracture surfaces. Strong evidence for H enhanced dislocation mobility has been obtained using *in situ* TEM studies, macroscopic stress strain curves, measurements of thermally activated dislocation motion, and theoretical treatments of hydrogen shielding of elastic stress centers.

Possible explanations for the "mixed" observations of hydrogen caused softening and hardening in macroscopic flow stress measurements have been advanced on the basis of damage caused by charging and slip localization. Hydrogen caused slip localization can result in increases in the macroscopic flow stress despite enhanced dislocation mobility.

REFERENCES

1. D. G. Westlake, *Trans. ASM 62* (1969) 1000.
2. H. K. Birnbaum, M. Grossbeck and S. Gahr, in I. M. Bernstein and A. W. Thompson (eds.), *Hydrogen in Metals,,* ASM Metals Park, Ohio, 1973, p. 303.
3. S. Gahr, M. L. Grossbeck and H. K. Birnbaum, *Acta Metall..,* 25 (1977) 1775.
4. D. Shih, I. M. Robertson and H. K. Birnbaum, *Acta Metall.,* 36 (1988) 111.
5. C. D. Beachem, *Metall. Trans.,* 3 (1972) 437.
6. H. K. Birnbaum, in R. P. Gangloff and M. B. Ives (eds), *Environment-Induced Cracking of Metals* , Houston, N.A.C.E. 1988, p. 21.
7. S. M. Meyers et al, *Reviews of Modern Physics, 64* (1992) 559.
8. E. Sirois, P. Sofronis, H. K. Birnbaum, in S. M. Bruemmer, et al. (eds.), *Fundamental Aspects of Stress Corrosion Cracking* , The Minerals, Metals and Materials Society, New York, 1992, p. 173.
9. H. K. Birnbaum, in N. Moody and A. W. Thompson (eds.), *Hydrogen Effects on Materials Behavior* , The Minerals, Metals and Materials Society, New York, 1990, p. 639.
10. E. A. Steigerwald, F. W. Schaller and A. R. Troiano, *Trans. Metall. Soc. AIME 218* (1960) 832.
11. R. A. Oriani and P. H. Josephic, *Acta Metall. 22* (1974) 1065.
12. D. H. Sherman, C. V. Owen and T. E. Scott, *Trans. AIME, 242* (1968) 1775.
13. M. L. Grossbeck and H. K. Birnbaum, *Acta Metall.,* 25 (1977) 125.
14. D. Hardie and P. McIntyre, *Metall. Trans.,* 4 (1973) 124.
15. N. E. Paton and J. C. Williams, in I. M. Bernstein and A. W. Thompson (eds.), *Hydrogen in Metals,,* ASM, Metals Park, Ohio, 1973, p. 409.
16. K. Nuttall, in A. W. Thompson and I. M. Bernstein (eds.), *Effects of Hydrogen on the Behavior of Materials* , TMS, New York, 1975, p. 441.
17. S. Takano and T. Suzuki, *Acta Metall.,* 22 (1974) 265.
18. T. B. Flannagan, N. B. Mason and H. K. Birnbaum, *Scripta Metall.,* 14 (1981) 109.
19. T. Tabata and H. K. Birnbaum, *Scripta Metall. 18* (1984) 231.

20. I. M. Robertson and H. K. Birnbaum, *Acta Metall. 34* (1986) 353.
21. G. Bond, I. M. Robertson and H. K. Birnbaum, *Acta Metall. 36* (1988) 2193.
22. G. Bond, I. M. Robertson and H. K. Birnbaum, *Acta Metall. 35* (1988) 2289.
23. P. Rozenak, I. M. Robertson and H. K. Birnbaum, *Acta Metall. 38* (1990) 2031.
24. T. Tabata and H. K. Birnbaum, *Scripta Metall. 187* (1984) 947.
25. J. Eastman, T. Matsumoto, N. Narita, F. Heubaum and H. K. Birnbaum, in I. M. Bernstein and A. W. Thompson (eds.), *Hydrogen in Metals,* , TMS, New York, 1980, p. 397.
26. M. S. Daw and M. I. Baskes, *Phys. Rev. B 29* (1984) 6443.
27. J. P. Hirth and J. R. Rice, *Metall. Trans.. A, 11* (1980) 1501.
28. J. P. Hirth, *Phil. Trans. Royal Soc. Lon., Ser. A 295* (1980) 139.
29. T. Matsumoto, J. Eastman and H. K. Birnbaum, *Scripta Metall., 15* (1981) 1033.
30. H. Z. Xiao, Ph.D. Thesis. University of Illinois (1993).
31. H. Hanninen, T. C. Lee, I. M. Robertson, and H. K. Birnbaum, Proc. of Int. Conf. on Corrosion - Deformation Interactions (1992) In press
32. H. Fukushima and H. K. Birnbaum, *Acta Metall. 32* (1984) 851.
33. D. Lassila and H. K. Birnbaum, *Acta Metall. 34* (1986) 1237.
34. D. Lassila and H. K. Birnbaum, *Acta Metall., 35* (1987) 1815.
35. I. M. Robertson, T. Tabata, W. Wei, F. Heubaum and H. K. Birnbaum, *Scripta Metall. 18* (1984) 841.
36. H. Matsui, H. Kimura and S. Moriya, *Mat. Sci. Eng. 40* (1979) 207.
37. S. Moriya, H. Matsui and H. Kimura, *Mat. Sci. Eng. 40* (1979) 217.
38. H. Matsui, H. Kimura and A. Kimura, *Mat. Sci. Eng. 40* (1979) 227.
39. F. Zeides, Ph.D. Thesis. University of Illinois (1986).
40. E. Sirois, Ph.D. Thesis. University of Illinois (1992).
41. D. G. Ulmer and C. J. Altstetter, *Acta Metall., 28* (1991) 1237
42. W. A. McInteer, A. W. Thompson and I. M. Bernstein, *Acta Metall., 28* (1980) 887
43. E. Sirois and H. K. Birnbaum, to be published.
44. J. P. Hirth and B. Carnahan, *Acta Metall. 26* (1978) 1795.
45. P. Sofronis and H. K. Birnbaum, *J. Mech. and Phys. of Solids*, submitted.
46. A. W. Cochardt, G. Schoeck, and H. Wiedersich, *Acta. Met 3* (1955) 533.
47. A. N. Stroh, *Adv. Phys. 6* (1957) 418.
48. A. Kimura and H. K. Birnbaum *Acta Metall. et Mater. 38* (1990) 1343.

ACKNOWLEDGMENTS

This work was supported by the Department of Energy under grant DEFG02-91ER45439.

FIGURE CAPTIONS

Fig. 1   Schematic showing the coordinates of the interacting dislocations and the hydrogen atmospheres. The shear stress, $d\tau_H$ is the shear stress due to the hydrogen atmosphere in the area dS located at position (r,ϕ). The extent of the outer radius of the H atmosphere, R, is determined by convergence of the full elastic solutions.

Fig. 2   Isoconcentration contours of the normalized hydrogen concentration, C/C_0, around an isolated edge dislocation. Calculations are for the parameters of Nb containing $C_0 = 0.1$ (H/M = 0.1) at 300 K. $C/C_0 = 6$ - - - - - - ; $C/C_0 = 2.25$ - · - · - · ; $C/C_0 = 1.5$ - - ·· - - ·· - - .

Fig. 3   Isoconcentration contours of the normalized hydrogen concentration, C/C_0, around two parallel edge dislocations having equal Burgers vectors of magnitude, b, on the same slip plane. Calculations are for the parameters of Nb containing $C_0 = 0.1$ (H/M = 0.1) at 300 K. a) dislocation distance 10b, b) dislocation distance 8b, c) dislocation distance 6b. $C/C_0 = 6$ - - - - - - ; $C/C_0 = 3$ - · - · - · - ; $C/C_0 = 1.5$ - - ·· - - ·· - - .

Fig. 4   Plots of the normalized shear stresses on dislocation 2, τ_H / μ due to H, τ_D / μ due to dislocation 1, and the net shear stress $(\tau_D + \tau_H) / \mu$ vs. normalized distance along the slip plane, l/b, for parallel edge dislocations having equal Burgers vectors. Calculations were carried out for the parameters characteristic of Nb at 300 K.   Hydrogen free material ———— ; H/M = 0.01 - · - · - · - ;   H/M = 0.1 - - - - - - .

Fig. 5   Plot of the interaction energy, W, between an edge dislocation and a C interstitial solute having a tetragonal axis along the [100] as a function of the normalized distance along the slip plane, x_1/b. The C solute is located at a distance $x_2/b = -0.505$. The calculation was carried out for the parameters of Nb at 300 K. H/M = 0 (pure Nb) ———— ; H/M = 0.1 — · — · — .

Fig. 6   Plot of the interaction energy, W, between an edge dislocation and a C interstitial solute having a tetragonal axis along the [010] as a function of the normalized distance along the slip plane, x_1/b. The C solute is located at a distance $x_2/b = -0.505$. The calculation was carried out for the parameters of Nb at 300 K. H/M = 0 (pure Nb) ———— ; H/M = 0.1 — · — · — .

Fig. 7   Plot of the interaction energy, W, between an edge dislocation and a C interstitial solute having a tetragonal axis along the [001] as a function of the normalized distance along the slip plane, x_1/b. The C solute is located at a distance $x_2/b = -0.505$. The calculation was carried out for the parameters of Nb at 300 K. H/M = 0 (pure Nb) ———— ; H/M = 0.1 — · — · — .

Fig. 8   Schematic diagram showing the effects of the elastic shielding by hydrogen at various temperatures and strain rates.

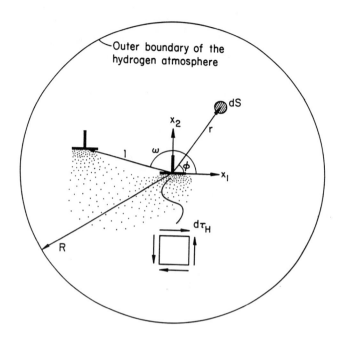

Fig. 1

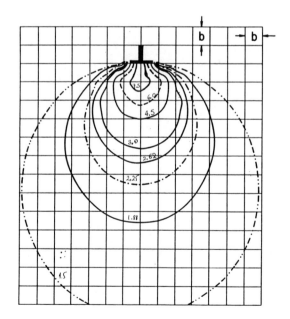

Fig. 2

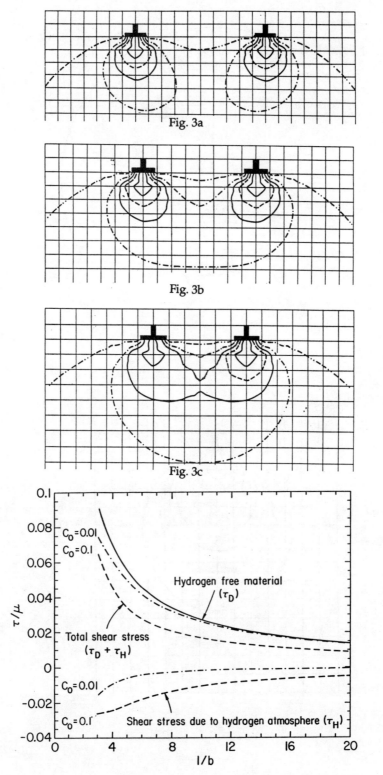

Fig. 3a

Fig. 3b

Fig. 3c

Fig. 4

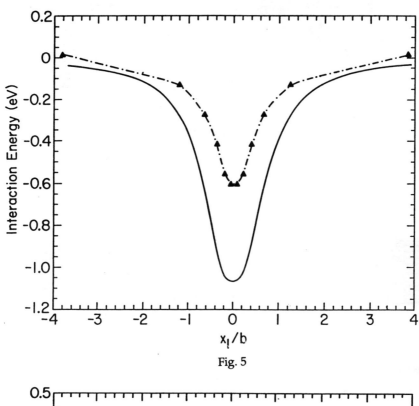

Fig. 5

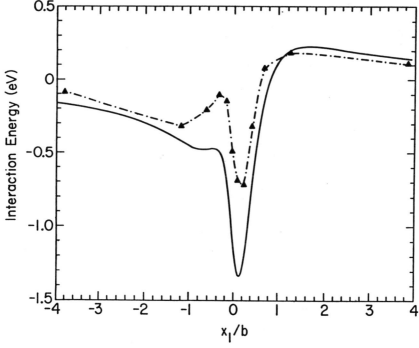

Fig. 6

29

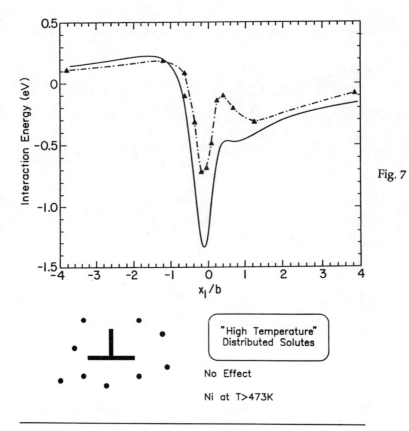

Fig. 7

"High Temperature"
Distributed Solutes

No Effect

Ni at T>473K

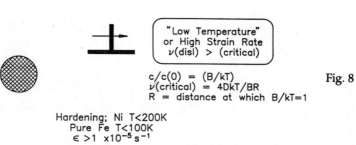

"Low Temperature"
or High Strain Rate
$\nu(disl) > (critical)$

$c/c(0) = (B/kT)$
$\nu(critical) = 4DkT/BR$
$R = $ distance at which $B/kT=1$

Fig. 8

Hardening; Ni T<200K
Pure Fe T<100K
$\dot{\epsilon} >1 \times 10^{-5} s^{-1}$

Intermediate Temperature
or Low Strain Rate
$B/kT > 1$

Ni 200 − 300K; $\dot{\epsilon}<1 \times 10^{-6}s^{-1}$
Fe 77 − 400K

High $\dot{\epsilon}$ −Hardening
Ni − 200K; $\dot{\epsilon}>1\times10^{-5} s^{-1}$

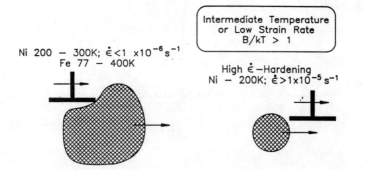

AD-Vol. 36, Fatigue and Fracture of Aerospace Structural Materials
ASME 1993

A FAILURE ASSESSMENT STUDY OF CASTING PARTS

W. L. Cheng
Corporate Technology Center

E. Bigelow
Ground Systems Division
FMC Corporation
San Jose, California

ABSTRACT

Allowable sizes of surface imperfections were determined for different regions of a casting that is subjected to cyclic loading. Finite element analysis was used to determine the nominal stresses in various sections of the casting. A novel approach based on fracture mechanics theory was applied to treat linear defects in the casting. Cumulative damage in sub-critical crack growth regime was determined using Paris Crack Growth relation. For standard cast material, a simplified fracture criterion was assumed to determine ultimate failure. At regions where surface hardening heat treatment was specified, ultimate failure occurred due to cleavage fracture and a brittle fracture criterion was applied. In the case of rounded defects in the castings, a conventional fatigue approach was used to determine their tolerance to cyclic loads. Results obtained from this study have provided a basis for determination of inspection criteria for both linear and rounded discontinuities.

1.0 INTRODUCTION

In recent years, the application of computer methods to stress, fatigue and fracture problems for failure analysis and life prediction has provided structural engineering community added tools for use in the design of complicated structures. One important application of this field of study pertains to controlling the surface quality of castings by identifying allowable surface imperfections that would not compromise part function.

This study applied a design approach which used global finite element modeling to identify nominal stress conditions and local modeling which utilizes fracture mechanics techniques to evaluate sub-critical crack growth and ultimate failure for linear defects, and conventional fatigue methodology to assess damage tolerance for rounded defects. Material background data for a specific cast component was determined for two regions: with standard quench and tempered heat treatment and with additional surface hardening heat treatment. For both types of defects, an iterative procedure was used to predict the life for crack growth through different sections based on initial defect size, section geometry, material properties, and cyclic stresses.

The tolerable sizes of linear and rounded defects were determined for different regions of the casting. Due to the ductility of the as-cast material, most parts of the casting exhibited stable crack growth after fatigue cracks have been initiated except for those areas with surface hardening heat treatment which demonstrated unstable crack propagation. The significance of establishing realistic inspection criteria requirements for nondestructive testing has the potential to increase component performance through design optimization, reduce quality control costs associated with scrap and rework, and minimize development costs associated with testing and redesign.

An engineering design approach was developed to establish criteria for evaluating both linear and rounded defects on the surface of a casting subjected to cyclic operating loading. Results were obtained using fracture mechanics theory and conventional fatigue approach which accounts for initiation and propagation for linear and rounded defects, respectively. The approach established for the linear defects is applicable to defects which may appear in the form of cracks, and cold shuts, hot tears or shrinkage in the casting. The one established for the rounded defects is applicable to defects which may appear in form of scabs, sand holes and porosities.

In this paper, the treatment of the rounded defects using the conventional fatigue approach will not be elaborated. Only a brief discussion will be offered. However, the fracture mechanics approach and its application to the failure assessment of linear defects will be discussed in details.

2.0 SURFACE QUALITY

Surface imperfections such as cracks, tears, holes and porosities can be detected using various nondestructive test methods including magnetic particle, eddy current, liquid-penetrant, and visual inspection. Excessively severe inspection criteria can lead to additional expense as a result of unnecessary scrap and rework if every defect must be addressed. Furthermore, inspection requirements must be efficiently directed at critical regions of the casting to allow foundries to focus on unsatisfactory trends and prevent introduction of imperfections that could affect performance. With cost and performance in mind, the initial consideration of what inspection criteria is required carries significant impact.

The results from this analysis demonstrate a practical method for evaluating surface imperfections in castings that are subjected to cyclic loading. Reference plots indicating maximum defect size versus applied stress range were characterized for different cross sections so that the allowable defect size in various regions of the casting could be determined providing that stress results were quantified using either conventional design calculations or the finite element method (FEM). These plots can also be applied to castings with similar material, section geometry and types of loading. In the past, surface quality requirements were established by more or less an art that focused on mixing some science with experience to obtain requirements for new designs or castings with quality problems. The approaches outlined in this paper allow wide spread use of fracture mechanics theory and conventional fatigue curves to casting design for the purpose of defining surface imperfection criteria by establishing requirements based on fundamental concepts for fatigue crack growth and material behavior.

3.0 APPROACH

The determination of defect tolerances for this casting consisted of two parts: stress analysis and cumulative damage analysis. The stress analysis for this casting was handled by using the finite element method because like most castings it was complicated in geometry. The tolerances for linear defects such as cracks, and cold shuts, hot tears or shrinkage were determined using fracture mechanics theory, and for rounded defects such as scabs, holes and porosities were determined using conventional fatigue methodology.

For linear defects, the fracture mechanics approach was justified as they usually have sharp notches which require considerably lower loads or fewer cycles than rounded defects before developing into macro fatigue cracks. These macro fatigue cracks have dimensions sufficient for their growth, and depend only on bulk properties, geometric and loading conditions rather than on local effects. This approach was different from typical strength of materials approaches. Conventional design procedures assumed the casting to be defect-free and a safe design would be achieved by making sure that the maximum stress was less than the ultimate stress (or yield stress) of the material divided by a factor of safety.

The fracture mechanics design methodology applied was based on the realistic assumption that all materials contain crack-like flaws. This analysis was performed on the casting with cracks placed in the most critical sites. A characteristic quantity defining the propensity of the crack to extend was determined. Usually, one primary crack was assumed for each critical site. The characteristic quantity applied (fracture parameter) depended on the particular failure criterion used, which may be the stress intensity factor, the strain energy release rate, the J-integral or the crack opening displacement. Selection of the appropriate quantity depended on the materials, loading and environmental conditions, and geometry of casting.

For rounded defects, the conventional fatigue approach was used as the total life of these defects constitutes the initiation and the propagation portions of the life. Details of this approach can be found in reference texts by Fuchs (1980) or Dowling (1993) and will not be discussed here. Essentially, the procedure encompasses the calculation of the local stress range and mean stress using local finite element analyses for a variety of rounded defects, adjustment of the local stress due to the mean stress using Goodman's diagram, and life prediction using fatigue damage analysis. For a fixed design life, an iterative process was used to determine the allowable size of rounded defects at different regions of the structure.

The following sections will provide an introduction to the stress analysis and various aspects of fracture mechanics that were applied to the failure assessment of linear defects.

3.1 Stress Analysis

Full knowledge of the cyclic stress distribution in this casting was subjected to operating loads is required to determine its defect tolerance. This could have been accomplished by either manual methods or finite element analysis depending on the complexity of the part, type of loading and boundary conditions. In practice, a global approach was adopted to determine the nominal principal stresses. Results from this analysis were used to determine the cyclic stress range and mean stress distributions over the entire casting. Effects due to stress concentrations were accounted for using empirical formulas, charts, or local finite element modeling. In the case of rounded defects, effects of the notch were captured using local finite element modeling approach.

3.2 Damage Tolerance Analysis

Linear defects in castings usually appear in the form of cracks, cold shuts, hot tears or shrinkage. The treatment of these defects was handled by fracture mechanics. In the domain of fracture mechanics, two distinct disciplines are usually discussed: brittle and ductile fracture. In most failure problems, both brittle and ductile fracture may occur during the failure process. There is usually a transition stage when both interact. Factors for determining the nature of the fracture (brittle or ductile) included intrinsic material behavior, geometry such as thickness and width, rate of loading, temperature and other environmental effects. When the material was brittle (as measured by the fracture toughness), it was likely to fail in a brittle manner for a thick section (i.e. plane strain condition). When the material was relatively ductile, failure of the material was either brittle or ductile depending on other factors such as plane strain or plane stress conditions. If the loading rate was high such as in impact loading (resulting in a one-time overload situation), brittle fracture occurred. This was similarly the case if the material was subjected to low temperature (below the ductile to brittle transition temperature).

This study deals with linear defects and takes into consideration crack propagation life and ultimate failure from ductile fracture for this particular cast material. Fracture mechanics based on the crack propagation relation originally formulated by Paris (1961) was used to determine the total life once a fatigue crack has initiated.

Two distinctively different approaches were exercised in treating these two types of fracture (brittle and ductile). Linear Elastic Fracture Mechanics (LEFM) has successfully been used to treat brittle fracture. This approach has a well-established use. Nonlinear Elastic Fracture Mechanics (NEFM) has been applied to solve problems involving ductile fracture. Some successes have been reported on specimen tests, but no well-defined universal methodology has yet been established for structural prediction. The area remains an active research topic and is evolving. The following sub-sections provide a more detailed description of the two topics.

3.2.1 Fracture Mechanics

Linear elastic fracture mechanics deals with the simplest fracture situation in which the material response was assumed linear and elastic. This was applies to the case of brittle fracture and was handled within the domain of linear

elastic fracture mechanics. Under this regime, the stress intensity factor, K_I, was defined by Irwin (1957) as a function of loading and geometry which characterized the elastic stress singularity associated with the crack tip. (1) shows the equation governing the opening mode (Mode I) and considers the stress intensity factor for general geometry, the shape of the crack, and the combined membrane and bending loads [Rooke, et al (1976), Murakami (1987)]:

$$K_I = (\sigma_t + H\sigma_b)\sqrt{\frac{\pi a}{Q}}F(\frac{a_{eff}}{t}, \frac{a_{eff}}{c}, \frac{a_{eff}}{b}, F) \quad (1)$$

where a_{eff} is the effective crack length

c is the crack depth

t, b are the thickness and width

σ_t = Tensile axial stress

σ_b = Bending stress

F = Angle around the crack front of the elliptical crack

H, Q, F are parameters [Newman et al (1978a, 1978b)]

Even though linear elastic material response was assumed, a small plastic region exists in reality near the crack tip at which a complex stress state prevails. High tri-axial normal stresses exist in this region The plastic zone size existing at the crack tip was included by using a plasticity correction factor given in (2a) and (2b) for plane strain and plane stress respectively:

$$r_p^* = \frac{K_I^2}{18\pi\sigma_{ys}^2} \qquad \text{for plane strain(2a)}$$

$$r_p^* = \frac{K_I^2}{6\pi\sigma_{ys}^2} \qquad \text{for plane stress(2b)}$$

where r_p^* is the plasticity correction factor, K_I is the stress intensity factor, and σ_{ys} is the yield stress. The effective crack length becomes

$$a_{eff} = a_o + r_p^* \qquad (2c)$$

where a_0 is the actual physical crack length

If the concern was fracture due to one-time loading such as in the case of an overload situation and if the material was brittle in nature, then a simplified brittle fracture criterion [Broek (1986), Atkins et al (1988)] as given in (3) could was employed to predict the initiation of unstable fracture propagation:

$$K_I \geq K_{IC} \text{ or } K_C \qquad (3)$$

where K_I = Stress Intensity Factor
　　　　　　　　at the Crack Tip

　　　　K_{IC} = Plane - Strain Mode I
　　　　　　　　Fracture Toughness

　　　　K_C = Plane - Stress Mode I
　　　　　　　　Fracture Toughness

The above criterion was a simplified fracture criterion based on only a single parameter. It suggests that unstable crack propagation has initiated when the above criterion was met. The criterion has been used for practical design of components susceptible to defects. More complicated criteria could be considered such as that using the fracture resistance-curve approach found in Creager (1973).

The above fracture criterion would break down for cases where the materials were ductile, substantial plastic deformation would have occurred before fracture initiated and propagated to ultimate failure. Brittle fracture criteria would no longer be adequate to provide an accurate prediction. In this case, ductile fracture criteria such as those based on J-Integral proposed by Rice (1968), Crack Tip Opening Displacement (CTOD) by Garwood (1985) or strip yield model by Dugdale (1960) should be used.

3.2.2 Fatigue Crack Propagation

In the case of fatigue, the life of the component comprises both crack initiation and propagation stages. An exact transition from initiation to propagation is usually not possible. A

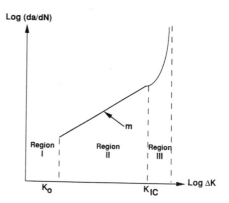

Figure 1　Characteristics of Fatigue Crack Growth Rate Curve da/dN - ΔK

typical fatigue crack growth rate curve is shown in Figure 1. This curve has the characteristic sigmoidal shape of a da/dN versus ΔK curve, where da/dN is the crack growth rate per cycle, and ΔK is the peak-to-peak range in the stress intensity factor K during the fatigue cycle, which divides the curve into three regions according to the curve shape, the mechanisms of crack extension and various influences on the curve. In region I there is a threshold value, ΔK_{th}, below which cracks do not propagate. Above this value the crack growth rate increases relatively rapidly with increasing ΔK. In region II there is often a linear log-log relation between da/dN and ΔK. Finally, in region III the crack growth rate curve rises to an asymptote where the maximum stress intensity factor, K_{max}, in the fatigue stress cycle becomes equal to the critical stress intensity factor, K_{IC} or K_C and catastrophic failure occurs.

Any point along the fatigue crack growth rate curve can be used as the fatigue crack growth limit and be used as a criterion for design of fatigue sensitive castings. The values of ΔK_{th}, and of da/dN in the near-threshold region, effectively control high-cycle fatigue behavior in most practical situations. This is because, in the first place, most real components and structures contain defects from which cracks may initiate relatively easily, making crack propagation the life-controlling process, and, secondly, because the nature of the growth rate dependence implies that a crack will grow relatively slowly in the initial stages, accelerating rapidly as its length increases.

In some cases, ΔK_{th} was selected as the limit for design of components and structures that did not contain linear defects, and materials used were not notch sensitive. In other cases where linear defects were present, the inclusion of fatigue crack growth rate beyond the threshold value was required.

There have been many attempts to describe the crack growth rate curve by "crack growth relation", which usually are semi or wholly empirical formulae fitted to a set of data. One of the most commonly used relations is the one proposed by Paris (1961) as shown in (4) in which he extended the use of the fracture parameter K to describe fatigue crack propagation in terms of the peak-to-peak range of K in the fatigue cycle, ΔK:

$$\frac{da}{dN} = C(\Delta K_I)^m \qquad (4)$$

where da/dN is the crack growth rate per cycle, C and m are constants which can be determined by tests. This simple empirical fit which provides a straight line on a log-log plot has been applied with remarkable success over the whole range of metallic materials, and in some cases to non-metallic materials.

Paris' equation describes only the linear log-log (region II) part of the crack growth curve, as indicated in Figure 1. Forman's equation [Forman (1967)] as described in (5) which also describes region III presents another approach.

$$\frac{da}{dN} = \frac{C(\Delta K_I)^m}{[(1-R)K_C - \Delta K_I]} \qquad (5)$$

where da/dN is the crack growth rate per cycle, C and m are constants which can be determined by tests, R is the load ratio which is equal to K_{min}/K_{max} as K_{min} being the minimum value of K_I in a regular fatigue cycle and K_{max} being the maximum value, K_C is the fracture toughness and ΔK is the peak-to-peak stress intensity range.

It is well understood that the crack growth rate curve is strongly dependent on the maximum stress, mean stress and the environment. It is recommended that the characterization of the curve be performed according to conditions similar to those the component sees.

In dealing with surface defects in castings, a pre-existing semi-elliptical crack was assumed. Due to various boundary conditions along the crack front, the stress intensity was not uniform. For example, in the case of membrane loading, as pointed out by Newman et al (1979a), the stress intensity factor was minimum at the end of the sharper curvature, i.e. near the surface, and maximum at the point where the curvature is gradual, i.e. at the bottom of the crack front. As a result, the crack growth rate was different along the crack front. In the case of semi-circular surface cracks, which is a special case of semi-elliptical surface cracks, the situation is similar. The crack growth of shallow flaws was under plane strain and therefore, was estimated by using the K-based crack growth approach such as the Paris or Forman relation. In general elliptical linear defects, such as part-through cracks, propagate in the thickness direction and become through cracks. Evolution of crack propagation as obtained by Nakamura et al (1986) is illustrated in Figure 2.

In this study, the length of the through crack was assumed to be that of the part-through crack at break-through. If the length of the part-through crack at break-through was less than the thickness (i.e. c< t) the through crack was taken equal to the thickness. This conversion was in accordance with

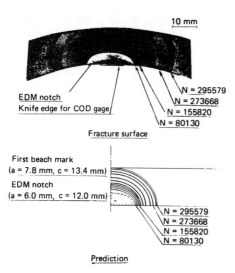

10 mm

EDM notch
Knife edge for COD gage

N = 295579
N = 273668
N = 155820
N = 80130

Fracture surface

First beach mark
(a = 7.8 mm, c = 13.4 mm)
EDM notch
(a = 6.0 mm, c = 12.0 mm)

N = 295579
N = 273668
N = 155820
N = 80130

Prediction

Figure 2   Crack Growth Evolution for a
Linear Surface Defect

general practice. If c<t, the crack converted to a circular shape during break-through, and then continued as a circular crack, so that c=t after the break-through.

When the applied stresses were sufficiently low, the maximum allowable crack size was such that it could represent a through-wall crack. Calculations of fatigue life for through wall cracks were based on a different crack-type and geometry function for the initial defect. As indicated later, there was a discontinuity along the curves for the maximum allowable defect size of the remaining ligament as defects got close to the thickness of the section.

There were cases in which the initial defects are deep part-through cracks, i.e. the crack depth was close to the thickness, then the accuracy of the fatigue calculations would deteriorate. This is especially true in the early stage of the propagation life, only a relatively small number of cycles are required to bring the part-through to a through-thickness crack. For all practical purposes, the calculations were performed as if it were a through crack.

3.2.3 Derivation of Maximum Allowable Defect Size

In order to determine the maximum allowable defect size, a pre-existing linear defect was assumed. The type of the defect considered in this study was a semi-circular surface crack which was common for this casting. Another assumption applied included that the crack had initiated to a fatigue crack so crack propagation was only a macro material phenomenon instead of a micro one. Under these assumptions, the shape, geometry and load functions were determined based on empirical formulae or numerical charts [Rooke et al (1976), Murakami (1987), Newman et al (1979a,b)] which were used to determine the stress intensity factor K_I for each updated crack geometry. The crack extension from the application of each load cycle was determined using the Paris Law. The crack size was updated until the final allowable crack size was reached based on either a brittle or a ductile fracture criterion depending on the cast material. The standard cast material (ASTM A148, Grade 105-85) exhibited moderate ductility. Ultimate failure is believed to be due to plastic collapse instead of unstable crack propagation, therefore a simplified failure criterion based on net section collapse was used. In this case, the average of the ultimate and yield strength of the material was selected as the collapsed stress. In regions where additional heat treatment (surface hardening) took place, the material exhibited brittle behavior and the K-based fracture criterion was used.

A computer program based on the above procedure was used to determine the fatigue life of the casting, which in turn was compared to its required life. If the fatigue life did not meet the required life, another initial crack size was tried until it was met. An iterative scheme was therefore developed to obtain the allowable defect size for each applied load and geometry. A set of applied loads was selected for this study, and based on those the maximum principal stress distribution was obtained using finite element analysis of the casting.

3.2.4 Allowable Defect Size Design Curves

A series of data in terms of the maximum allowable defect size versus applied stress have been calculated by the above procedure. The data were developed using ASTM standard fracture test from laboratory test specimens [Werner (1989)]. Plots of allowable defect size versus stress range were obtained and used for design application. Each plot represents different geometries of a critical region on the casting. The geometry was defined by the effective width and thickness of the critical section. The crack shape was assumed semi-circular with dimensions of a (2c) being the crack length and c being the depth. Figure 3 shows a typical design curve for a specific cross section. The loading applied included membrane loading (σ_t in Equation 1) as well as bending (σ_b in Equation 1) in which the gradient effects are taken into account. In those cases, an effective stress was used resulting in an effective K value.

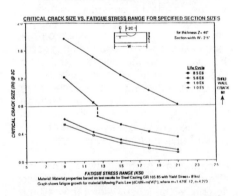

CRITICAL CRACK SIZE VS. FATIGUE STRESS RANGE FOR SPECIFIED SECTION SIZES

Figure 3 A Typical Plot of Crack Size versus Applied Stress Range

There were two distinct sections in most of the plots. One section of the curve represents the high-stress range in which the maximum allowable defects were part-through cracks while the other section of the curve shows the low-stress range in which the maximum allowable defects were through-thickness cracks. As mentioned in the previous section, it would take relatively few cycles or lower loads for a deep crack to form a through-thickness crack. For all practical purposes, these were considered through cracks to begin with. Due to the limited number of discrete points, it was decided that the low-stress section of the curve would be extrapolated to the line drawn horizontally representing the through-thickness dimension. From that point, a vertical line was drawn downward to meet the high-stress section of the curve. At any point where there were multiple design values, the lower bound value was used. This approach eliminated the danger for users to select a non-conservative maximum allowable defect size for designs.

3.2.5 Application of the Design Curves

Using these design data in determining the defect tolerance of castings required complete knowledge of the stress state over the entire casting. In essence, the maximum allowable defect size was determined for each of the critical regions throughout the castings depending on the stress distribution.

When a single maximum allowable defect size was of interest such as in the case when defect found at a certain location of the part, then only the stress state at that location was required. The characterization of the stress distribution could be accomplished by manual method, finite element analysis or test depending on the complexity of the part, type of loading and boundary condition. In practice, a global approach was adopted to determine the nominal stresses. The effect due to stress concentrations could be accounted for using empirical formulae or charts, or local finite element modeling.

Once the stresses are known, the section geometry for critical locations of the component was defined. Based on geometry and applied stress, the maximum allowable defect size was determined from the appropriate plots. When the section geometry was different from those given in the plots or the applied stresses were not within the range studied, new data was required. In some cases, engineering judgement was used with the existing data even though it does not exactly match the problem.

This procedure does not mention the treatment of safety factor explicitly. It was assumed that a safety factor was included in the load specified by the design requirement. If safety factor were to be used, it may be applied to the stresses used in the calculations of the defect sizes.

4.0 EXAMPLE

The above methodology was applied to define the maximum linear defects allowable in various critical regions of a cast sprocket. The nominal stress range distribution for the casting under operating loads was determined by using the finite element method. Due to the complexity of the part, a three-dimensional model was required. The model was developed using a general-purpose finite element modeling package - IDEAS. Due to symmetry, only half of the casting was modeled (Figure 4). The model consisted of about five thousand 20-node isoparametric brick elements with bubble function. A sinusoidal load distribution was applied to the outside rim of the casting to simulate a pulling load in addition to a constant torque. Restraints were applied at the center to simulate the mounting of this component onto a rigid plate with bolts. Several stress analyses were performed using the linear static option of the MSC/NASTRAN finite element program. The first analysis simulated an actual static test to determine the quality of the finite element model. Good correlation was obtained providing confidence in the model. Subsequent analyses simulated the operational load condition and established the stress range

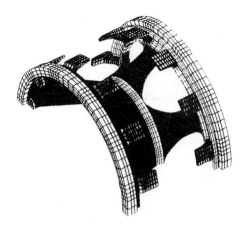

Figure 4 Finite Element Model of a Complicated Casting

distribution. Figure 5 shows the stress distribution of the combined loads.

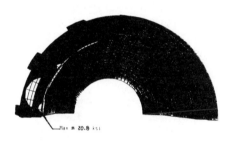

Figure 5 Stress Distribution of a Complicated Casting due to Operating Loads

Several critical regions of the casting were considered and were labeled as Regions 1 through 6. Their minimum dimensions were defined in Table 1. The cyclic stresses in terms of maximum principal stresses were used as defects may lie at any orientation and the maximum principal stresses (only tensile) were the most critical ones to cause failure. The corresponding maximum principal stress range in these regions were also shown in this same table.

Table 1 Minimum Dimensions and Stress Range at Critical Sections of the Casting

Critical Section	Minimum Thickness (inches)	Minimum Width (inches)	Stress Range (ksi)

(Material: ASTM A148 Grade 105-85, Q & T, 105 ksi Min. TS, 85 ksi Min. YS, 17% Min. elongation, 35% Min. reduction in area)

Region 1	0.50	2.88	19.2
Region 2	0.38	7.93	20.8
Region 3	0.34	6.52	6.20
Region 4	0.72	4.96	4.90
Region 5	0.39	1.47	9.00

(Material: GR 105-85, Induction hardened to Rockwell C-47 to C-55, 3% elongation & 5% Min. reduction in area)

Region 6	0.30	1.91	12.4

In the case of linear defects, a surface correction factor was used on the principal stress range at locations where quarter-circular cracks were possible. By using the Paris crack growth relation, the fatigue damage was accumulated cycle by cycle until reaching a converged solution in the iterative process. The maximum crack size was determined for each region and the results were summarized in Table 2.

Table 2 Calculated Allowable Linear and Rounded Defect Size at Critically Stressed Sections of a Grade 105-85 Steel Casting

Critical Section	Linear Defect Size (inches)	Rounded Defect Size (inches)
Region 1	0.19	0.32
Region 2	0.20	0.20
Region 3	thru-wall	thru-wall
Region 4	thru-wall	thru-wall
Region 5	0.15	thru-wall
Region 6	0.15	0.17

It should be noted that a ductile fracture criterion based on plastic collapsed was used for Regions 1 through 5. In Region 6 however, a brittle fracture criterion was required because the material was brittle as a result of heat treatment (induction hardening). In addition to steady state cyclic loading, a sudden overload situation was also considered to determining the allowable defect size in this region. The lesser of the two defect sizes was selected for the acceptable design value.

In the case of rounded defects, critical notch stresses including the range and the mean value at each rounded defect were determined using local finite element analysis. A correction on the stress range due to mean stress was made using Goodman's diagram. Fatigue life was determined using conventional S-N approach. Using an iterative procedure, the maximum rounded defect size was determined for each region and the results were also summarized in Table 2.

It should be mentioned that a plastic stress concentration factor was used in the calculation of the corrected stress range for defects at Regions 1 through 5 as the as-cast material behaved in a ductile manner versus a conventional stress concentration factor based on elastic stress for defects at Region 6.

As expected, the tolerable defect size is more generous for rounded defects. This is especially the case for the as-cast material. Only a slightly larger defect size was allowed for the rounded defects than the linear ones in the induction-harden section due to the brittleness of the material.

5.0 CONCLUSIONS

Conventional fatigue and fracture mechanics based approaches were applied to establish acceptance criteria for both rounded and linear defects in a casting subjected to cyclic operating loads. This approach took into account different sections of the casting and included material changes due to surface hardening heat treatment. The approach enabled us to specify the defect size allowed in different regions of the casting without affecting the part function. A typical casting was selected to demonstrate the application of the approach. The criteria used to establish the defect size in critically stressed regions provided adequate fatigue life, and relaxed tolerance in non-critically stressed regions. This ultimately maintained part function while eliminating unnecessary rework and reducing part reject rates.

ACKNOWLEDGMENT

The authors appreciate numerous suggestions provided by M. Mineart during the early phase of this project. Some of the analyses were performed by S. Doan during his employment with FMC as a consultant.

REFERENCES

Atkins, A.G.; and Mai, Y.W., 1988, Elastic and Plastic Fracture, Halsted Press, John Wiley & Sons.

Broek, D., 1986, Elementary Engineering Fracture Mechanics, 4th Revised Edition, Martinus Nijhoff Publishers.

Creager, M., 1973, "A Note on the Use of a Simple Technique for Failure Prediction Using Resistance Curves," STP 527, ASTM.

Dowling, N.E., 1993, Mechanical Behavior of Materials, Prentice-Hall International Editions.

Dugdale, D.S., 1960, " Yielding in Steel Sheets Containing Slits," J. Mech. Phys. Solids, 8.

Forman, R.G.; Kearney, V.E.; and Engle, R.M., 1967, " Numerical Anslysis of Crack Propagation in Cyclic-Loaded Structures," J. Basic Engineering.

Fuchs, H.O., and Stephens, R.I., 1980, Metal Fatigue in Engineering, Wiley-Interscience Publication.

Garwood, S.J., 1985, " A Crack Tip Opening Displacement (CTOD) Method for the Analysis of Ductile Materials," Eighteenth National Symposium on Fracture Mechanics, Boulder, Colorado, USA.

Irwin, G.R., 1957, "Analysis of Stresses and Strains Near the End of a Crack Traversing a Plate,", Journal of Applied Mechanics, 24.

Nakamura, H.; Matsushima, E.; and Okamoto, A., 1986 , " The Influence of Residual Stress on Fatigue Cracking," PVP-Vol. 103, ASME.

Murakami, Y., 1987, Stress Intensity Factors Handbook, Pergamon Press.

Newman, J.C. Jr.; and Raju, I.S., 1979a, "An Empirical Stress-Intensity Factor Equation for The Surface Crack," Engineering Fracture Mechanics, Vol. 15, Pergamon Press.

Newman, J.C. Jr.; and Raju, I.S., 1979b, "Stress-Intensity Factors for a Wide Range of Semi-Elliptical Surface Cracks in Finite-Thickness Plates," Engineering Fracture Mechanics, Vol. 11, Pergamon Press.

Paris, P.C.; Gomez, M.; and Anderson, W.E., 1961, " A Rational Analytic Theory of Fatigue," in Trends in Engineering, 13.

Rice, J.R.; and Rosengren, G.F., 1968, "Plane Strain Deformation Near a Crack Tip in a Power-Law Hardening Material," J. Mech. Phys. Solids, 16.

Rooke, D.P.; and Cartwright, D.J., 1976, Compendium of Stress Intensity Factors, Ministry of Defense, Her Majesty's Stationery Office, London.

Werner, A., 1989, " Fatigue and Fracture Properties Determination of Grade 105-85 Cast Steel," MEL Report No. 891630, FMC, Corporate Technology Center.

AD-Vol. 36, Fatigue and Fracture of Aerospace Structural Materials
ASME 1993

ACCEPTANCE CRITERIA FOR IN-SERVICE INSPECTION OF HEAT EXCHANGER HEAD AND SHELL COMPONENTS

P. S. Lam, R. L. Sindelar, and N. G. Awadalla
Savannah River Technology Center
Westinghouse Savannah River Company
Aiken, South Carolina

ABSTRACT

Several heat exchanger structural components of the Savannah River Site (SRS) production reactors, comprising the pressure boundary for the primary or secondary coolant systems, are monitored through in-service inspection. An acceptance criteria methodology is provided to disposition flaws detected during examination of these components. The methodology ensures that the safety margins against failure through flaw instability are maintained throughout service.
Acceptance criteria are established to define the acceptable flaw configurations. The acceptable flaw lengths are estimated based on the flaw stability solutions for throughwall cracks. To limit the flaw depths for leakage prevention, the acceptable depths are set to 75% of the plate thicknesses for flaws with lengths up to the acceptable flaw length.
The stresses acting on the heat exchanger components were obtained by simple analytical and finite element solutions. The stability of postulated flaws were evaluated with elastic-plastic fracture mechanics and limit load analyses. Curvature correction methods were employed to approximate solutions for flaws on curved shells.

INTRODUCTION

The code-of-record for the SRS K-reactor Process Water System (PWS) replacement heat exchangers is Section VIII, Division 1 (no stamp) of the American Society of Mechanical Engineers (ASME) Boiler and Pressure Vessel (BPV) Code (1980 & 1983 Editions). In-service inspection of the heat exchangers is not required per the construction code. The service history of the heat exchanger shows that service-induced degradation has led to several instances of cracking and leakage failures in the heads and shells. Improvements in the design and materials of construction of the heat exchangers have been made throughout the reactor operating history to reduce or eliminate the conditions that have led to these failures. In-service inspections of these external pressure boundary components provide a monitoring function of the heat exchangers and can show that service-induced degradation that led to cracking has been eliminated or that the degree of degradation is acceptable.

The acceptance criteria for the reactor PWS heat exchanger heads, shells, and cooling water nozzles have been developed in this paper to disposition flaws, both circumferential and axial, identified and characterized in the in-service inspections. The acceptance of flaws is in accordance with the ASME BPV Code, Section XI (1992) to maintain the code-based factors of safety on loading specified by IWB-3642 for acceptance-by-analysis in austenitic stainless steel piping.

The development of the acceptable flaw configurations includes a stress analysis of the components with the factor of safety of 3 on normal and 1.5 on emergency/faulted (off-normal) loads. The residual stresses are also included in the fracture mechanics analysis. Fracture mechanics analysis is performed to determine stability of postulated throughwall flaws at various orientations and locations in the component. A limit on part-throughwall surface flaws to 75% of the thickness of the component, in accordance with IWB-3641 and Code Case N-494 (1992), is imposed to preclude inadvertent leakage. This depth criterion, together with the most limiting length for throughwall flaws under normal and off-normal conditions, define the acceptable flaw configuration for each pressure boundary component.

The mechanical properties applied in the structural and fracture mechanics analyses are listed for the heat exchanger pressure boundary components. The stability of postulated flaws in the head, shell, and cooling water nozzles is performed with elastic-plastic fracture mechanics and limit load analyses. Curvature corrections are used to approximate solutions for flaws in curved shells. The acceptance criteria methodology for demonstrating the structural integrity is summarized with a flow chart.

SRS HEAT EXCHANGER

Each of the six primary coolant loop systems of the SRS production reactors contains two parallel single-pass heat exchangers to transfer heat from the moderator or primary coolant (D_2O) to the secondary cooling water (H_2O). The heat exchangers are horizontal, saddle-mounted cylindrical tanks (bounded by the shell and tubesheets) about 33.5 feet long and 7.5 feet in diameter (Fig.1). The moderator enters the heat exchanger head-tubesheet plenum space and is distributed to about 9000, 1/2" OD and 0.049" nominal thickness tubes. The tubes pass through the shell (350.5" long) and are supported by baffles and the tubesheets which are attached at two ends to the heat exchanger inlet and outlet heads by 84 staybolts, the primary load-carrying components of the head. The secondary system cooling water (river water) enters through an inlet nozzle in the bottom of the shell at the PWS outlet end of the heat exchanger and exits through an outlet nozzle in the top of the shell at the PWS inlet end. The heat exchangers are mounted on two rail trucks with seismic bracing to anchor them firmly to the floor.

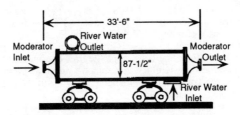

Figure 1. General View of SRS Production
Reactor Heat Exchanger

HEAT EXCHANGER COMPONENT MATERIALS

Tensile Properties

All the SRS K-reactor heat exchanger heads, shells, and cooling nozzles are fabricated from wrought Type 304L stainless steel. This material is resistant to sensitization and Intergranular Stress Corrosion Cracking (IGSCC), and is therefore superior to the Type 304 stainless steel materials used in the earlier SRS heat exchanger designs. Tensile properties for Type 304L stainless steel are from Section II of the ASME BPV Code (1992) and from SRS-specific data. The flow stress, σ_f, applied in limit load analyses of the piping (Stoner, 1991) is taken as $3S_m$ where S_m is the stress intensity value from Section II of the code. The stress intensity value for Type 304L plate is 16.7 ksi and the flow stress is 50.1 ksi for temperatures from -20 to 300°F. The modulus of elasticity (Young's modulus), E, at a nominal temperature of 200°F is given in ASME BPV Code as 27,600 ksi. The value of E at 200°F is conservative in predicting the material tolerance of flaws at temperatures lower than 200°F which bounds the normal and off-normal operating conditions. A value of 0.3 is taken as Poisson's ratio.

Ramberg-Osgood parameters for the tensile response are applied in the elastic-plastic fracture analysis of Type 304L stainless steel material. Results from tensile testing of archival materials are applied in this analysis. The Ramberg-Osgood parameters with values of $\alpha=2$ and $n=5$ were derived

(Stoner, 1991) for test data from Type 304 (Stoner, 1991) and 304L (Hawthorne, 1987) stainless steel specimens of identical design.

Fracture Toughness

Mechanical test data from a reference plate of Type 304L stainless steel provides the elastic-plastic fracture toughness of the heat exchanger materials. Two compact tension specimens of Type 304L plate material with a 0.4T planform design (Stoner, 1991) were tested at 257°F (Hawthorne, 1987). The average of the parameters for the power law fit to the J-R curve results give the following for the material J-R curve (deformation J):

$$J \left[\frac{lb}{in}\right] = 9370(\Delta a)^{0.3847}$$

with Δa, the crack extension, in inches. The limit on Δa for J-controlled crack extension is suggested to be 0.1181 inches (Stoner, 1991). The corresponding limit of J at Δa of 0.1181 inches is 4120 lb/in.

The tearing modulus (T) is

$$T = \frac{E}{s_f^2} \frac{dJ}{da}$$

where s_f is defined as the average of the yield and tensile (engineering) strengths. With the average flow stress data at 257°F from tensile specimens, the material tearing modulus can be formulated as:

$$T = \frac{42.21}{\Delta a^{0.6153}}$$

with Δa in inches. The power law formulation for $J(\Delta a)$ along with the tearing modulus, $T(\Delta a)$, define the material J-T curve for the fracture mechanics analysis.

FRACTURE ANALYSIS OVERVIEW

Fracture analysis was performed for postulated throughwall flaws in the heat exchanger shell, cooling water inlet nozzle, and head. A safety factor of 3 (for normal and design conditions) or 1.5 (for off-normal or emergency/faulted conditions) was applied to the calculated stresses at the corresponding loading conditions for input to the fracture analysis to determine instability flaw lengths. The acceptable flaw length for each component is the shortest of the instability lengths or the most limiting length from the flaw stability results of the different fracture analysis methods and loading cases.

Limit load and J-integral methods for pipes or cylinders were applied to postulated axial and circumferential cracks in the shell part of the heat exchanger. The J-integral based limit load analysis (JiLLA), developed for SRS piping (Awadalla, 1987), was applied to analyze postulated flaws at the circumferential weld of the cooling water inlet nozzle and in the moderator inlet region of the heat exchanger head. The postulated flaws in the remaining part of the head were analyzed by the J-integral method.

A finite element model for the heat exchanger head was used to calculate the stress distribution in the head including seismic loads in the off-normal case. A double curvature correction was developed for the J-integral solution in the highly-curved region of the head.

COMPONENT 1: HEAT EXCHANGER SHELL

The heat exchanger shell is a cylinder with mean radius (R) of 44" and thickness (t) of 0.5". The following loading conditions are considered in the calculation of throughwall flaw instability lengths for the heat exchanger shell:
  (1) Design pressure: 150 psi (normal operating pressure is 50 psi at cooling water inlet).
  (2) A bending moment of 4,820,000 in-lb due to the weight of heat exchanger (251,000 lbs).
  (3) Water hammer pressure at partial check valve closure: 182 psia (168 psig).
  (4) Water hammer pressure at full check valve closure: 256 psia (242 psig).
  (5) Seismic response: The longitudinal force is 17,900 lb and the bending moment is 1,790,000 in-lb. The cooling water pressure is assumed at normal condition.
  (6)Heat exchanger tube rupture: Shell may be exposed to process water design pressure of 300 psi.

45

The hoop and longitudinal stresses (σ_H and σ_L, respectively) due to pressure loading (P) are estimated according to:

$\sigma_H = \dfrac{PR}{t}$ and $\sigma_L = \dfrac{PR}{2t}$. The maximum bending stress (σ_b) is $\dfrac{MR_o}{I}$, where R_o is the outside radius of the shell, M is the applied bending moment, I is the moment of inertia of the shell (133,811 in^4) and t is the thickness of the shell.

In the evaluation of flaw stability, a safety factor of 3 is applied to the design condition. For faulted conditions such as water hammer, seismic and tube rupture events, the safety factor 1.5 is applied to the resultant stresses.

Limit Load Analysis

The limit load analysis was carried out separately for axial flaws and for circumferential flaws. The contribution from residual stress is not applied in the limit load analysis since the method is based on net section plastic collapse whereby the residual stresses would have been relieved.

Axial Cracks. The axial crack extension criterion for high-toughness vessels was proposed by Hahn et al. (1969) as

$\sigma_f = M \, \sigma_H.$

The factor $M = \sqrt{1 + 1.61 \, \lambda^2}$ was used by Eiber (1971), where $\lambda = a/\sqrt{Rt}$ and a is the half crack length. The acceptable flaw length can be obtained by solving for a from these equations under various loading conditions described earlier. Only the hoop stress due to internal pressure was considered as the applied load for longitudinal cracks.

Circumferential Cracks. The applied loads include the longitudinal stress (σ_L) due to internal pressure of the shell, and the bending stress (σ_b) due to the weight of heat exchanger and the seismic loads. In this case, the maximum tensile stress is used (i.e., $\sigma = \sigma_L + \sigma_b$). Therefore, the acceptable flaw length is calculated based on the solution for a circumferential crack under tension (Zahoor, 1989):

$$\frac{\sigma}{\sigma_f} = \frac{1}{\pi} [\, 2 \cos^{-1} (\tfrac{1}{2} \sin \theta) - \theta]$$,

where θ is the half crack angle ($\theta = a/R$).

Elastic-plastic J-integral Analysis (J-T Method)

The J-integrals were calculated for cracks in axial and circumferential directions. The crack growth ($J \geq J_{IC}$) is stable if $T < T_R$, where T_R is the tearing modulus of the material. In practice, the intersection point of the applied J-T curve and the material J-T curve will determine the crack growth stability limit (Mehta, 1989). In the present case, a cut-off J value (4120 lb/in) obtained in material testing is used to determine the instability crack length.

The elastic-plastic solutions for J-integral in a pipe or cylinder are not available for the size of heat exchanger shell (R/t=88). A solution of J-integral for a center-cracked panel (CCP) under plane stress condition for Ramberg-Osgood materials was obtained by Shih and Hutchinson (1976). This analytical solution provides a basis to construct an approximate solution for the geometry of the heat exchanger shell by a curvature correction factor. The validity of this correction was evaluated by comparison to a finite element solution for the case of circumferential cracks as discussed in the Appendix.

The procedure of applying the J-T methodology to postulated flaws in the heat exchanger shell is summarized in the following:

(1) For a given applied remote stress, calculate the CCP solution of Shih and Hutchinson for various crack lengths. The J-integral (J^{ccp}) is composed of an elastic portion (J_{el}^{ccp}) and a plastic portion (J_{pl}^{ccp}), that is,

$$J^{ccp} = J_{el}^{ccp} + J_{pl}^{ccp}.$$

(2) The plastic zone size correction (or small scale yielding correction) is applied to J_{el}^{ccp}.

(3) For a given crack length, calculate the ratio (Y) of the stress intensity factor for a cylinder with the same R/t to $\sigma\sqrt{\pi a}$, the stress intensity factor for an infinite flat plate.

(4) Analogous to a compounding method for approximating stress intensity factors, and noting that J is proportional to the square of the stress intensity factor in linear elastic fracture mechanics, the corrected values of elastic and plastic components of J for the heat exchanger shell (J_{el}^{cur} and J_{pl}^{cur}) are

$$J_{el}^{cur} = Y^2 \, J_{el}^{ccp} \text{ and}$$

$$J_{pl}^{cur} = Y^2 \, J_{pl}^{ccp}, \text{ respectively.}$$

(5) The contributions of fracture parameters from other sources, such as thermal stress or residual stress, can be combined in the sense of linear elastic fracture mechanics. The elastic portion of J-integral (J_{el}^{cur}) in (4) above is first converted to K_I^{appl}, the Mode I stress intensity factor due to applied loads.

$$K_I^{appl} = \sqrt{E \, J_{el}^{cur}}, \text{ for the plane stress condition.}$$

(6) The stress intensity factor due to the residual stress (K_I^{res}) is evaluated. The residual stress distribution is a self-equilibrium, symmetric pattern with maximum tension ($+\sigma_r$) on the edges and maximum compression ($-\sigma_r$) in the mid-section of the plate. The through-thickness variation from tension-compression-tension is linear. A formula based on finite element solutions was obtained by Green and Knowles (1992):

$$K_{max}^{res} = 0.43\sigma_r\sqrt{\pi t}$$

Note that the K_I^{res} is saturated to a maximum value when the crack is extended in length only a fraction of the plate thickness. Therefore, the residual stress of this type is not subject to curvature correction.

(7) The total elastic portion of J is calculated as

$$J_{el} = \frac{1}{E}\left(K_I^{appl} + K_I^{res}\right)^2.$$

(8) The plastic portion of J remains unchanged, that is, $J_{pl} = J_{pl}^{cur}$.

(9) The total J-integral of the crack in the heat exchanger shell is

$$J = J_{el} + J_{pl}.$$

The expressions for J-integral in CCP and the curvature correction factors (Y^2) are summarized in the Appendix.

COMPONENT 2: COOLING WATER INLET NOZZLE

The cooling water nozzle is a 3/8" thick pipe circumferentially welded to the heat exchanger shell. A J-integral based limit load analysis (JiLLA) is adopted to evaluate the acceptable flaw length for the cooling water inlet nozzle.

The normal operating pressure for cooling water inlet is 50 psi (the design pressure is 150 psi). The nozzle loads were given as 10,100 lb in its axial direction and a resultant bending moment of 575,000 in-lb. These loads include seismic, thermal, and gravity contributions. Adding a longitudinal stress due to the normal operating pressure, the total applied membrane stress in the axial direction becomes 1126 psi. The maximum bending stress is calculated to be 3552 psi. These applied stresses with a safety factor of 1.5 are used in the fracture analysis.

J-Integral Based Limit Load Analysis (JiLLA)

The results in the SRS reactor process water piping leak-before-break analysis (Awadalla, 1987; Mehta, 1987) showed that the theoretical limit load result for a circumferential crack under bending could accurately predict the maximum loads in experiments, especially for small diameter pipes (e.g., 4" diameter piping). However, the limit load solutions do not contain explicit information of pipe diameter and wall thickness; the acceptable flaw size based on a given load may be over-estimated for large pipe sizes. On the other hand, the J-T approach with General Electric or Electric Power Research Institute (EPRI) estimation scheme does contain the pipe size information, but the measured maximum loads in experiments are underpredicted.

To use the limit load solution and to maintain the pipe size information, an empirical factor (M') was developed to scale the limit load results for pipes under bending. The M' factor is the ratio of the instability stress (predicted by J-T approach) for a 4" diameter, Schedule 80 pipe and that for another pipe of larger size (e.g., 24" diameter, Schedule 20). In the case of heat exchanger cooling water nozzles, the value of M' is 1.6. Therefore, for a given circumferential crack, the modified instability stress would be the maximum stress predicted by a limit load analysis divided by M'.

The procedure of using JiLLA concept to estimate the acceptable circumferential flaw length for cooling water inlet nozzle weldment under significant bending load is as following:

(1) Apply a safety factor (SF) to the applied longitudinal membrane stress (σ_m^{appl}) and the bending stress (σ_b^{appl}).

$$\sigma_m = SF \times \sigma_m^{appl}$$

$$\sigma_b = SF \times \sigma_b^{appl}$$

(2) Scale the maximum stress by factor M' with the membrane stress unchanged.

$$\sigma_m + \sigma_b^{'} = M' \left(\sigma_m + \sigma_b\right)$$

(3) Calculate the modified bending stress ($\sigma_b^{'}$) in terms of the applied stresses.

$$\sigma_b^{'} = SF\left[M'\left(\sigma_m^{appl} + \sigma_b^{appl}\right) - \sigma_m^{appl}\right]$$

(4) Solve for half crack angle $\left(\theta = \dfrac{a}{R}\right)$ from the limit load equation for the bending capacity of a throughwall circumferential crack:

$$\sigma_b^{'} = \frac{2\,\sigma_f}{\pi}\left(2\,\sin\beta - \sin\theta\right)$$

where

$$\beta = \frac{\pi}{2}\left(1 - \frac{\theta}{\pi} - \frac{\sigma_m}{\sigma_f}\right)$$ defines the location of the neutral axis.

COMPONENT 3: HEAT EXCHANGER HEAD

The primary restraint of the head pressure boundary is provided by 84 staybolts attached to the tubesheet. The head can be divided into three general regions regarding to its shape (Figure 1): the flat region including the head flange (Region I), the highly-curved region (Region II), and the 12-inch pipe region (Region III) which connects the process water piping system (moderator inlet). The diameter of the head including the flange is 96.25" and its height is 26.125". A thickness of 1.5" was used in stress analyses.

The normal operating pressure and the pressure surge during the pump startup are bounded by the design pressure of 300 psi. In addition, piping reaction forces due to seismic, thermal, and gravity conditions can occur at the pipe flange. The loads include a force of 10,500 lb in the heat exchanger axial direction, forces of 4500 lb and 6800 lb in the lateral directions, a torsion of 182,000 in-lb along the axial direction, and moments of 420,000 in-lb and 168,000 in-lb acting about two remaining Cartesian coordinate axes. In addition, the reaction force due to the process water design pressure (300 psi) in the 12-inch pipe is also added to the axial force, resulting in 44,429 lb in the axial direction.

Maximum principal stresses are calculated through a finite element analysis of the heat exchanger head loaded by the hydrostatic design pressure of 300 psi and the set of piping reaction forces. It was noted that the pipe flange seismic loading has insignificant impact on the stress distribution of the head, in the areas away from the pipe region, when all 84 staybolts are intact. It implies that the finite element results, obtained with faulted stress boundary conditions, are also valid for the design condition. A safety factor of 3 is applied to the maximum principal stresses in Regions I and II for the flaw stability analysis.

For the pipe region of the head (Region III), the J-integral based limit load analysis (JiLLA) is adopted. A safety factor of 1.5 is applied to the axial force (44,429 lb) and the resultant bending moment (452,400 in-lb) in accordance with the JiLLA approach.

Finite Element Analysis

The finite element model (Lam, 1993) contains 1224 shell elements including 216 thick shell elements representing the pipe flange and the head flange and 300 beam elements for the 84 staybolts. The heat exchanger head is loaded with 300 psi design pressure and the piping reaction forces. It was found that the highest value for the maximum principal stress in the curved region is about 6 ksi; and in the flat region is about 11 ksi. These stresses are used in the fracture analysis with J-T method to determine the acceptable flaw lengths in the highly-curved region (Region II) and in the flat region (Region I) of the heat exchanger head.

Elastic-Plastic J-integral Analysis (J-T Method)

The concept and procedure for J-T method applied to the head of heat exchanger is similar to those described for the heat exchanger shell. In the highly-curved region, a double curvature correction empirical procedure is developed to account for the effects of meridional and tangential radii of curvature (Appendix). In the flat region, the base solution is directly employed without curvature corrections. The residual stress distribution is assumed to have the same pattern as in the case of heat exchanger shell. The stress intensity factor due to the residual stress is calculated in the same manner as in the case of heat exchanger shell discussed earlier.

HEAT EXCHANGER SHELL AND HEAD ACCEPTANCE CRITERIA

The acceptance criteria for flaws (produced by service-induced degradation) in the external pressure boundary of the heat exchanger are the acceptable flaw length and the acceptable depth of 75% of the thickness of the respective components. The resultant flaw lengths maintain the ASME-based margins of safety in Section XI (1992), Article IWB-3642 for austenitic stainless steel piping. The fracture analysis was performed for postulated throughwall flaws and is, therefore, more limiting than part-throughwall flaws at the same length. Flaw depth up to 75% of the piping wall thickness is provided by Section XI, Article IWB-3641 (1992). Note that the 75% depth requirement was also adopted in ASME BPV Code Case N-494 (1992).

The following Tables provide the acceptable flaw lengths for the head, shell, and cooling water nozzles by identifying the shortest instability lengths or most limiting lengths from the results obtained by various methods described earlier.

TABLE 1
Acceptable Flaw Lengths in Heat Exchanger Head

Region	Maximum Principal Stress (ksi)	Safety Factor	Acceptable Flaw Length (Inches)
I. Flat (incl. Flange)	11	3	26
II. Highly-curved	6	3	16 (meridional)
III. Pipe	(not applicable)	1.5	23 (circumferential)

Note: The dominant flaw directions are identified next to the flaw length. In the flat region the analysis does not distinguish preferential directions. The flaw lengths for Regions I and II were determined with J-T method. The flaw length for Regions III was determined with the J-integral based limit load analysis (JiLLA).

TABLE 2
Acceptable Flaw Lengths in Heat Exchanger Shell

Crack Orientation	Most Limiting Case	Analysis Method	Acceptable Flaw length (inches)
Axial (Longitudinal)	P=150 psi Design Pressure Safety Factor = 3 or P=300 psi Tube Rupture Safety Factor = 1.5	Limit Load Approach	6
Circumferential	P=150 psi Design Pressure Safety Factor = 3	J-T Approach	30

TABLE 3
Acceptable Flaw Lengths in Heat Exchanger Cooling Water Nozzle

Crack Orientation	Most Limiting Case	Analysis Method	Acceptable Flaw length (inches)
Circumferential	Nozzle loads and normal operating pressure 50 psi Safety Factor = 1.5	JiLLA	39

ACCEPTANCE CRITERIA METHODOLOGY FOR HEAT EXCHANGER PRESSURE BOUNDARY COMPONENTS

Basis for In-Service Inspection

Leakage failures have occurred in the service history of the heat exchangers. Examinations (with visual, surface, or volumetric methods), conducted under an in-service inspection program, provide information on the service condition of the component and also provide the information showing that service-induced degradation has been mitigated or eliminated. An acceptance criteria methodology is a framework for periodically monitoring the service condition through in-service examinations and includes the disposition of degraded conditions while maintaining safety margins against failure. For heat exchanger pressure boundary components, it involves the following sequence in an in-service inspection program: 1) Baseline Examination; 2) Evaluation; 3) Acceptance; and 4) Successive Examinations.

Acceptance Criteria Methodology

The acceptance criteria methodology for the pressure boundary components of the heat exchangers is shown in Figure 2. Baseline examinations of the heat exchanger components would be performed in accordance with the code case for In-Service Inspection for Low Temperature Heavy Water Reactors, presently in draft (Cowfer, 1992). Evaluation and acceptance of reported flaws would include comparison of the flaw size with the allowable sizes in the acceptance standards for the heat exchanger components provided in the code case. If the flaw size exceeds the allowable sizes in the acceptance standards, then acceptance-by-analysis (e.g., comparison of the size to the results of this present analysis and consideration of growth rate) and/or additional examinations would be required for acceptance of the condition. If degradation is significant, periodic monitoring through successive examinations would be continued with an increase in the frequency of examinations.

For example, if an relevant indication does not exceed the allowable flaw size specified in the acceptance standards, no further evaluation would be necessary and examination of the component would be performed at the next interval in the inspection program. If a flaw exceeds the size in the acceptance standards, acceptance-by-analysis and/or additional examinations would be required per the code case for acceptance of the flaw. Evaluation of the service-induced degradation and the flaw growth rate would be performed, and, if a flaw is predicted to exceed the acceptance criteria (developed in the present analysis) prior to the next interval in the inspection program, then the flaw would be re-inspected before the next interval such that the acceptance criteria are met. The flaw growth rate evaluation would be updated following each successive examination or reexamination.

The acceptance criteria provided in this paper are the acceptable flaw configurations (length and depth) to maintain the defined safety margins for the heat exchanger head, shell, and cooling water nozzles serving as the limit to flaw configurations acceptable for service. No flaws in the pressure boundary components of the heat exchanger shall exceed or be predicted to exceed these configurations. Flaws in the pressure boundary of the heat exchanger that exceed or are predicted to grow to exceed the acceptable configurations while in service would require repair or replacement of the heat exchanger.

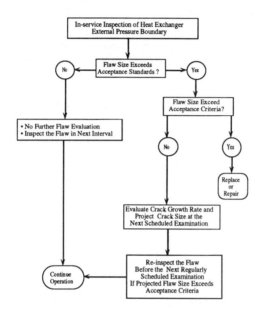

Figure 2. Flow Chart for Disposition of Flaws

ACKNOWLEDGMENT

The information contained in this article was developed during the course of work done under Contract No. DE-AC09-89SR18035 with the U.S. Department of Energy.

REFERENCES

ASME Boiler and Pressure Vessel Code, Section II, "Materials," 1992 Edition.

ASME Boiler and Pressure Vessel Code, Section VIII, Division 1, 1980 and 1983 Editions.

ASME Boiler and Pressure Vessel Code, Section XI, "Rules for Inservice Inspection of Nuclear Power Plant Components," 1992 Edition.

ASME Boiler and Pressure Vessel Code, 1992 Code Cases, Nuclear Components, N-494: "Pipe Specific Evaluation Procedures and Acceptance Criteria for Flaws in Class 1 Ferritic Piping that Exceed the Acceptance Standards of IWB-3514.2, Section XI, Division 1," (approved December 3, 1990), p. 829.

Awadalla, N.G., Sindelar, R.L., Caskey, G.R., and Abramczyk, G.A., 1987, "Reactor Materials Program: Leak-Before-Break Evaluation of Weldments Reactor Process Water Piping" DPST-87-837, Savannah River Laboratory, E.I. du Pont de Nemours & Co., Aiken, SC.

Cowfer, C.D., 1992, "Requirements for Inservice Inspection of Heavy Water Production Reactors (U)," WSRC-TR-90-42-141, Westinghouse Savannah River Co., Aiken, SC [initial draft of the code case being prepared by the ASME SCXI Special Working Group - Low Temperature Heavy Water Reactors].

Eiber, R.J., Maxey, W.A., Duffy, A.R., and Atterbury, T.J., 1971, "Investigation of the Initiation and Extent of Ductile Pipe Rupture," BMI-1908, Battelle Columbus Laboratories, Columbus, OH.

Erdogan, F., 1982, "Theoretical and Experimental Study of Fracture in Pipelines Containing Circumferential Flaws," DOT-RSPA-DMA-50/83/3, Department of Mechanical Engineering and Mechanics, Lehigh University, Bethlehem, PA, prepared for US Department of Transportation.

Green, D. and Knowles, J., 1992, "The Treatment of Residual Stress in Fracture Assessment of Pressure Vessels," in Proceedings of 1992 Pressure Vessels and Piping Conference, New Orleans, LA, PVP-Vol.233, Pressure Vessel Fracture, Fatigue, and Life Management, American Society of Mechanical Engineers, pp.237-247.

Hahn, G.T., Sarrate, M., and Rosenfield, A.R., 1969, "Criteria for Crack Extension in Cylindrical Pressure Vessels," International Journal of Fracture Mechanics, Vol. 5, pp.187-210.

Hawthorne, J.R. and Hiser, A.L., 1987, "Sample Preparation, Irradiation, and Testing of 304 Stainless Steel Specimens," MEA-2219, Materials Testing Associates, Inc., Lanham, MD.

Lam, P.S., Sindelar, R.L., Barnes, D.M., and Awadalla, N.G., 1993, "CTOD-Based Acceptance Criteria for Heat Exchanger Head Staybolts," in the Proceedings of the 2nd ASME/JSME Nuclear Engineering Joint Conference, Vol.2, P.F. Peterson edi., pp.107-116.

Mehta, H.S. and White, M.A., 1987, "Application of Leak-Before-Break Approach to SRP Process Piping" SASR#87-24 (DRF137-0010), General Electric Nuclear Energy, San Jose, CA.

Mehta, H.S., 1989, "Fracture Mechanics Evaluation of Potential Flaw Indications in the Savannah River L, P and K Tanks," SASR#86-64 (DRF137-0010), General Electric Nuclear Energy, San Jose, CA.

Rooke, D.P. and Cartwright, D.J., 1976, Compendium of Stress Intensity Factors, Her Majesty's Stationery Office, London.

Shih, C.F. and Hutchinson, J.W., 1976, "Fully Plastic Solutions and Large Scale Yielding Estimates for Plane Stress Crack Problems," Trans. American Society of Mechanical Engineers, Journal of Engineering Materials and Technology, Series H, Vol. 98, pp.289-295.

Stoner, K.J., Sindelar, R.L., and Caskey, Jr., G.R., 1991, "Reactor Materials Program - Baseline Material Property Handbook - Mechanical Properties of 1950's Vintage Stainless Steel Weldment Components (U)," WSRC-TR-91-10, Westinghouse Savannah River Co., Aiken, SC.

Tada, H., Paris, P.C., and Irwin, G.R., 1985, The Stress Analysis of Cracks Handbook, Second Edition, Pages 33.3, 33.4, 33.6 and 34.1, Paris Productions Incorporated (and Del Research Corporation), Saint Louis, MO.

Zahoor, A., 1989, Ductile Fracture Handbook, Volume 1, Pages 3-5 and 3-1, Electric Power Research Institute, Palo Alto, CA.

APPENDIX: ENGINEERING APPROACH FOR J-INTEGRAL APPROXIMATIONS FOR CRACKS IN A CURVED SHELL

Finite Plate Solution of Shih & Hutchinson (1976)

The J-integral solution for a crack in a Center-Cracked Panel (CCP) is used to construct approximate J-integrals for the postulated cracks in the heat exchanger shell. Both the axial and the circumferential cracks can be treated in a similar manner prior to curvature correction. The correspondence between the width of CCP and the shell can be seen in Figures A1 and A2. The procedure for calculating J-integral in a finite CCP is described in the following:

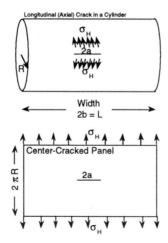

Figure A1. Longitudinal Crack and Corresponding Configuration of a Center-Cracked Panel

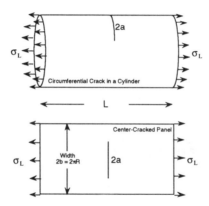

Figure A2. Circumferential Crack and Corresponding Configuration of a Center-Cracked Panel

The J-integral solution for a Ramberg-Osgood material is (Shih, 1976)

$$\frac{J}{\sigma_o \varepsilon_o a(1-a/b)} = \psi \left(\frac{P}{P_o}\right)^2 g_1\left(\frac{a_{eff}}{b}, n=1\right) + \alpha\left(\frac{P}{P_o}\right)^{n+1} g_1\left(\frac{a}{b}, n\right).$$

The elastic and plastic portions of J-integral are

$$J_{el}^{ccp} = \psi \, \sigma_o \varepsilon_o \, a(1-a/b) \left(\frac{P}{P_o}\right)^2 g_1\left(\frac{a_{eff}}{b}, n=1\right)$$

and

$$J_{pl}^{ccp} = \alpha \, \sigma_o \varepsilon_o \, a(1-a/b)\left(\frac{P}{P_o}\right)^{n+1} g_1\left(\frac{a}{b}, n\right),$$

where a is the half crack length, b is the half specimen width (Fig.A1 or A2), σ_o is set to $3S_m$ or 50.1 ksi for 304L stainless steel,

$\varepsilon_o = \sigma_o / E$,

$a_{eff} = a + \varphi\, r_y$, $P \leq P_o$ [†]

$a_{eff} = (a_{eff})_{P=P_o}$, $P > P_o$,

$\varphi = \dfrac{1}{1 + (P/P_o)^2}$,

$P_o = 2(b-a)\sigma_o$ is the lower bound limit load,

$P = 2b\sigma^\infty$ is the applied load corresponding to a remote stress σ^∞,

$r_y = \dfrac{1}{2\pi}\left(\dfrac{n-1}{n+1}\right)\left(\dfrac{K_I}{\sigma_o}\right)^2 = \dfrac{a}{2\pi}\left(\dfrac{n-1}{n+1}\right)\left(1 - \dfrac{a}{b}\right)\left(\dfrac{P}{P_o}\right)^2 g_1\left(\dfrac{a}{b},\,1\right)$ for plane stress, and

$\psi = \dfrac{a_{eff}}{a}\left(\dfrac{b-a}{b-a_{eff}}\right)$.

The function g_1 is calculated according to

$$G(a/b,\,n) = g_1(a/b,\,n)\left[\dfrac{1 + \zeta(n-1)a/b}{c\sqrt{n}\,(1 - 1/n) + g_1(a/b,\,1)/n}\right].$$

The best choice of the constants was found by Shih and Hutchinson (1976), that is, $\zeta = 1.40$ and $c = 3.85$. The function $g_1(a/b,\,1)$ is known from the elastic solution:

$$g_1(a/b,\,1) = \pi\left[1 - 0.5a/b - 0.370(a/b)^2 - 0.044(a/b)^3\right]^2.$$

For 304L stainless steel, Ramberg-Osgood exponent n is 5. The values of the function $G(a/b,\,n)$ are tabulated in Shih & Hutchinson and reproduced here.

a/b	G(a/b, n=5)
0	1.005
1/8	1.031
1/4	1.037
1/2	0.939
3/4	0.860
1	0.800

Interpolation was used to calculate $G(a/b,\,n=5)$ for values of a/b from 0 to 1.

Infinite Plate Solution of Shih & Hutchinson (1976)

In the case of an infinite plate, $a/b = 0$. The equations for a finite plate are greatly simplified.

The elastic and plastic portions of J-integral in an infinite CCP are denoted by $J_{el}^{\infty ccp}$ and $J_{pl}^{\infty ccp}$.

(i) The elastic portion of J-integral is calculated according to

$$J_{el}^{\infty ccp} = \dfrac{1}{E}\left(K_I^\infty\right)^2$$

[†] This equation is a modification of the one that originally appeared in Shih and Hutchinson (1976) by introducing a parameter φ which was proposed in "Fully Plastic Crack Solutions, Estimation Scheme, and Stability Analyses for the Compact Specimen," by V. Kumar and C. F. Shih, in Fracture Mechanics: Twelfth Conference, ASTM STP 700, American Society for Testing and Materials, 1980, pp. 406-438.

where

$$K_I^\infty = \sigma^\infty \sqrt{\pi a_{eff}}$$

and

$$\frac{a_{eff}}{a} = 1 + \frac{1}{2}\frac{n-1}{n+1}\frac{(\sigma_\infty/\sigma_o)^2}{1 + (\sigma_\infty/\sigma_o)^2} \qquad \text{for } \sigma_\infty \le \sigma_o$$

or

$$\frac{a_{eff}}{a} = 1 + \frac{1}{4}\frac{n-1}{n+1} \qquad \text{for } \sigma_\infty > \sigma_o.$$

Note that the above two equations vary slightly from those in Shih & Hutchinson (1976) due to the factor φ in the plastic zone correction.

(ii) The plastic portion of J-integral is

$$J_{pl}^{\infty ccp} = \alpha \sigma_o \varepsilon_o a \left(\frac{\sigma_\infty}{\sigma_o}\right)^{n+1} g_1(0,n)$$

where

$$g_1(0,n) \cong 3.85 \sqrt{n}\,(1 - 1/n) + \pi/n\,.$$

(iii) The total J-integral (J) for the CCP is, therefore,

$$J = J_{el}^{\infty ccp} + J_{pl}^{\infty ccp}$$

Curvature Corrections

The size of the heat exchanger shell prohibits direct use of existing J-integral solutions found in the literature. A curvature correction is applied to estimate the J-integrals. In linear elastic fracture mechanics, the stress intensity factor (K) for a crack can usually be expressed in a general form of $K = Y \sigma\sqrt{\pi a}$, where σ is the applied stress and Y ($Y = Y_1$ for axial crack and $Y = Y_2$ for circumferential crack) is a function of crack size and specimen dimensions. The curvature correction factor, used for estimating J-integral of a crack in a curved structure based on a corresponding flat plate solution with the same crack length, is therefore Y^2. The degree of validity of this procedure is demonstrated later in this Appendix for circumferential throughwall cracks.

The curvature correction factors for the heat exchanger shell or a pipe can be obtained with handbook solutions, such as Erdogen (1982) and Rooke and Cartwright (1976). The solutions provided by Tada et al. (1985) are simpler to use and the solutions are better behaved in some instances.

Axial (Longitudinal) Cracks. The stress intensity factor of an axial crack with length 2a subjected to a hoop stress σ_H in a cylinder with mean radius R and thickness t is (Tada, 1985)

$$K_I = \sigma_H \sqrt{\pi a}\; Y_1(\lambda),$$

where $\lambda = \dfrac{a}{\sqrt{Rt}}$,

$$Y_1(\lambda) = \sqrt{1 + 1.25\,\lambda^2} \quad \text{for } 0 < \lambda \le 1,$$

and

$$Y_1(\lambda) = 0.6 + 0.9\,\lambda, \quad \text{for } 1 \le \lambda \le 5.$$

In this case the curvature correction factor for J-integral for an axial or a longitudinal crack is Y_1^2.

Circumferential Cracks. For a circumferential crack with length 2a or angle 2θ subjected to a longitudinal stress σ_L, the stress intensity factors, based on Tada (1985) are

$$K_I = \sigma_L \sqrt{\pi a}\ Y_2(\lambda\ \text{or}\ \theta),$$

where

$$Y_2(\lambda) = \sqrt{1 + 0.3225\ \lambda^2}\ \text{ for }\ 0 < \lambda \leq 1,$$

$$Y_2(\lambda) = 0.9 + 0.25\ \lambda,\ \text{ for }\ 1 \leq \lambda \leq 5,$$

and

$$Y_2(\theta) = \sqrt{\frac{2}{\eta}\frac{1}{\pi\theta}}\ f(\theta),\ \text{ for }\ \lambda > 5.$$

In the last equation,

$$\eta^2 = \frac{t/R}{\sqrt{12\left(1 - v^2\right)}}\ \text{ and }\ f(\theta) = \theta + \frac{1 - \theta\cot\theta}{2\cot\theta + \sqrt{2}\cot\left(\frac{\pi - \theta}{\sqrt{2}}\right)}.$$

Therefore, the curvature correction factor for J-integral for a circumferential crack is Y_2^2.

Double Curvature Correction

The shape of the heat exchanger head at any location is defined by a pair of local radii of curvature: one is on the meridional plane containing the heat exchanger axis (R_m), the other is on the plane perpendicular to the heat exchanger axis (R_t, the radius from the heat exchanger centerline to the tangent line at that point on the head surface). In the highly-curved region of the head, the curvature effects on the fracture parameters are significant. A procedure for double curvature correction is developed for this region. A single pair of averaged radii of curvature is used for the correction. The averaged meridional radius of curvature (R_m) and the averaged tangential radius of curvature (R_t) for the highly-curved region are estimated to be 18.2" and 13.4", respectively.

The orientation of a crack in the curved region yielding the shortest instability length is in the meridional or axial direction of the heat exchanger head. This is due to the hoop stress as a result of process water pressure. For a crack in a structure with double curvature, such as in a bend pipe shown in Figure A3, this configuration can be achieved by first transforming a CCP to an axial crack in a cylinder through an operation Y_1^2, then by another transformation Y_2^2 into a circumferential crack.

Therefore, the final configuration is considered as a combined transformation, $Y_2^2 \otimes Y_1^2$, from the original configuration of a center-cracked panel. In an engineering approach for estimating J-integral for a crack in a structure with double curvature, the approximated solution is given as

$$J^{cur}_{el\,/\,pl} = Y_2^2\,Y_1^2\,J^{ccp}_{el\,/\,pl}$$

where $J_{el\,/\,pl}$ represents the elastic or plastic portion of J-integral, and Y_1 and Y_2 have been defined earlier for axial and circumferential crack curvature corrections, respectively.

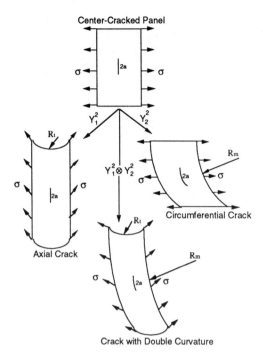

Figure A3. Transformation of a Center-Cracked Panel to
a Crack with Double Curvature

Finite Element Verification of Curvature Correction

The curvature correction method assumes that the proportionality factor (Y^2) obtained from the elastic portion of J also exists for the plastic portion. In an elastically dominant system, this treatment should provide an accurate approximation for the J-integral.

The degree of accuracy of curvature correction for the case of circumferential cracks under tension has been demonstrated by an elastic-plastic finite element analysis. Five finite element meshes[tt] have been constructed with the half crack angles including $\pi/16$, $\pi/8$, $\pi/4$, 0.4π and $\pi/2$. A typical mesh is shown in Figure A4.

Both the finite element analysis and the curvature correction procedure were performed for a cylinder (or pipe) with the same size as the heat exchanger shell (R/t=88), and for a Ramberg-Osgood material with α=2 and n=5. Three representative internal pressures (150, 300, and 450 psi) or associated longitudinal (axial) stresses were used to generate Figure A5. Note that the curvature correction curves contain discontinuities, because the stress intensity factor solutions used to derive the curvature correction factors contain a discontinuity at $a = 5\sqrt{Rt} = 23.5"$ (Tada, 1985). It can be seen in Figure A5 that the results with the curvature correction applied to a flat plate (CCP) solution nearly matches the finite element results, especially under the stress level of interest and for this class of circumferential cracks in the Ramberg-Osgood material (α=2 and n=5).

[tt] Courtesy of the South Carolina Universities Research and Education Foundation (SCUREF) Task No.68, "Advanced Fracture Mechanics to Assess Complicated Piping Flaws," by Mr. W. Gi, Dr. Y.-J. Chao, and Dr. M. A. Sutton, Department of Mechanical Engineering, University of South Carolina, Columbia, South Carolina, 1992.

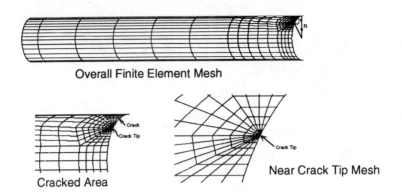

Overall Finite Element Mesh

Cracked Area

Near Crack Tip Mesh

Figure A4. Finite Element Mesh for a Circumferential Crack ($\theta = \pi/4$)
in a Cylinder with R = 44"

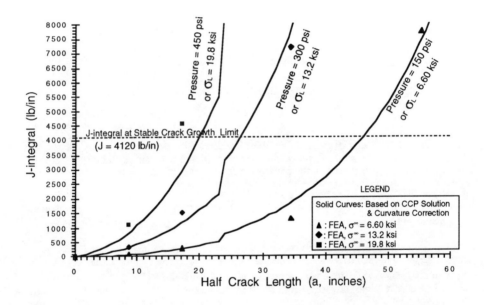

Figure A5. Comparison of J-integral Values from Direct Finite Element Calculations
and from Curvature Correction based on a Flat Plate Solution

AD-Vol. 36, Fatigue and Fracture of Aerospace Structural Materials
ASME 1993

LIFE PREDICTION IN HIGH TEMPERATURE
STRUCTURAL MATERIALS

Carl E. Jaske
CC Technologies, Incorporated
Columbus, Ohio

ABSTRACT

Life prediction methods are required to assess the performance and safety of the structural materials used in engineering systems and components that operate at high temperatures. At high temperatures, materials are subject to time-dependent creep and environmental degradation, as well as cycle-dependent fatigue degradation. The life prediction methods must account for all of these degradation mechanisms and their possible interactions.

The purpose of this paper is to review the methods that are used to predict the creep and fatigue life of structural materials. Traditional methods that have been used to predict the life of structural materials are based on the initiation of a significant crack; whereas more recent methods employ fracture mechanics to predict life based on the growth of a crack from some initial size to a critical size. Crack-growth based life prediction methods are emphasized in this review.

INTRODUCTION

Engineering components and systems that operate at elevated temperatures are subjected to both cyclic and steady loadings. Cyclic loadings can induce fatigue damage in structural materials, while steady loadings can induce creep damage in them. A combination of both cyclic and steady loadings can induce creep-fatigue damage interaction in structural materials. Thus, a general methodology for life prediction of structural materials must account for fatigue damage, creep damage, and creep-fatigue damage interaction.

Traditional life prediction methods are based on empirical crack-initiation models. Fatigue damage is accounted for using strain range versus fatigue life relationships. For example, the Manson (1953) - Coffin (1954) expression relates fatigue life, N_f, to plastic strain range, $\Delta\varepsilon_p$, as follows:

$$\Delta\varepsilon_p = A \, N_f{}^a \, , \qquad (1)$$

where A and a are empirically determined constants that depend on material and temperature. In a similar fashion, the Basquin (1910) expression relates fatigue life to the elastic strain range, $\Delta\varepsilon_e$, in the following manner:

$$\Delta\varepsilon_e = BA\, N_f{}^{ba} , \qquad (2)$$

where BA and ba also are empirically determined constants that depend on material and temperature. Equations (1) and (2) often are combined into a single expression that relates fatigue life to total strain range, $\Delta\varepsilon_t$, as follows:

$$\Delta\varepsilon_t = BA\, N_f{}^{ba} + A\, N_f{}^{a} . \qquad (3)$$

The American Society of Mechanical Engineers (ASME) Boiler and Pressure Vessel Code uses the Langer (1962) equation, which adds a fatigue limit term, $\Delta\varepsilon_L$, to Equation (1) in the following manner:

$$\Delta\varepsilon_t = A\, N_f{}^{a} + \Delta\varepsilon_L . \qquad (4)$$

In actual use, both sides of Equation (4) are multiplied by one-half the value of the elastic modulus to yield an expression relating stress amplitude to fatigue life. For inelastic straining, the stress amplitude is a fictitious value greater than the actual stress amplitude, but it is directly comparable with elastically calculated stresses (Anon., 1969), which simplifies its use in fatigue design applications.

Creep-rupture life, t_r, typically is related to stress, σ, and absolute temperature, T, using a time-temperature parameter (Conway, 1969). The most widely used of these is the well-known Larson-Miller (1952) parameter, LMP, which is defined below:

$$LMP = T\,(C + \log t_r) . \qquad (5)$$

C is the Larson-Miller constant and usually has a numerical value near 20. For a specific structural material, stress is some function of the Larson-Miller parameter as follows:

$$\sigma = f\{LMP\} = f\{T\,(C + \log t_r)\} . \qquad (6)$$

Often, the specific form of Equation (6) is a polynomial relating $\log \sigma$ to LMP, where the values of the coefficients in the polynomial are determined by least-squares regression analysis of data from creep-rupture tests.

A number of procedures has been developed to account for creep-fatigue interaction (Coffin, et al., 1977). Examples of these procedures are the linear life-fraction rule (Taira, 1962), the strain-range partitioning method (Manson, et al, 1971), and the frequency separation approach (Coffin, 1976). The linear life-fraction rule is simply a combination of Miner's (1945) cycle-fraction rule for fatigue damage and Robinson's (1952) time-fraction rule for creep damage, where the total material damage, D, is given by the following expression:

$$D = \Sigma\,(n_a/N_f)_i + \Sigma\,(t/t_r)_i , \qquad (7)$$

where n_a and N_f are the number of applied cycles and the number of cycles to failure, respectively, for each strain range in the loading history, and t and t_r are the time and time to rupture, respectively, for each stress level in the loading history. The left-hand term of Equation (7) is fatigue damage, whereas the right-hand term is creep damage. Usually, failure is predicted to occur when the value of D equals 1.

The linear life-fraction rule is used in the elevated-temperature design rules of ASME Code Case N-47. In this case, N_f in Equation (7) is replaced with the allowed number of cycles, and t_r in Equation (7) is replaced with the allowed rupture life. Thus, as employed in Code Case N-47, both damage fractions incorporate safety factors. Furthermore, a material-dependent interaction diagram is used to define allowed values of D and limit its value to much less than one in regimes where large amounts of creep-fatigue interaction have been observed in laboratory tests.

The strain-range partitioning approach extends the Manson-Coffin relation by partitioning inelastic strain range into the following four components: (1) fully reversed plastic straining, $\Delta\varepsilon_{pp}$, (2) plastic straining reversed by creep straining, $\Delta\varepsilon_{pc}$, (3) creep straining reversed by plastic straining, $\Delta\varepsilon_{cp}$, and (4) fully reversed creep straining, $\Delta\varepsilon_{cc}$. A cyclic-life relation similar to Equation (1) then is developed for each of these four types of strain ranges, so a total of four such equations are required to describe the cyclic life behavior of a material. As reviewed by Viswanathan (1989), a fractional inelastic strain damage procedure is to compute cumulative damage and predict cyclic life.

The frequency separation approach modifies the basic Manson-Coffin relation by incorporating a correction term to account for the effect of frequency on cyclic life. The term differentiates between tension-going and compression-going straining to account for the effects of creep-fatigue interaction on cyclic life. Application of the frequency separation approach also has been reviewed by Viswanathan (1989).

The above life-prediction methods are based on crack initiation or the fracture of small (about 6 mm diameter) laboratory specimens and do not explicitly address crack-propagation life. For many important and practical engineering structures, the major portion of material life is involved in the propagation of a crack from an existing defect or the limitations of nondestructive inspection require the conservative assumption that material life consists entirely of crack propagation from the largest defect that may not have been detected during inspection. Linear elastic fracture mechanics (LEFM) has been successfully applied to the prediction of crack-growth life in structural materials used at temperatures below the creep range. For high temperatures, where creep deformation can occur, it is usually necessary to employ inelastic fracture mechanics (IFM) as well as LEFM, especially for materials that exhibit ductile creep behavior. The remainder of this paper reviews the principles of IFM and describes the application of IFM and LEFM to the prediction of crack-growth life in high-temperature structural materials.

MODELING HIGH-TEMPERATURE CRACK GROWTH

The overall approach for modeling creep-fatigue crack growth that is reviewed in this paper is based on the model first proposed by Jaske and Begley (1978). A crack-tip-zone interaction model (Jaske, 1983a) was developed to account for creep-fatigue interaction effects. The approach has been applied to creep-fatigue crack growth in Type 316 stainless steel (Jaske, 1983b and 1984). It also has been used to predict crack-growth life in 2-1/4Cr-1Mo steel and 1-1/4Cr-1/2Mo steel (Jaske, 1990), and in this paper it is employed to assess creep-fatigue crack growth in 9Cr-1Mo-V-Nb steel.

Fig. 1 illustrates the concepts involved in predicting the crack-growth life of structural materials. The defect size distribution is characterized using nondestructive inspection or from destructive examinations of comparable materials. Physical damage, such as creep voids, and metallurgical damage, such as the formation of precipitates, is characterized using metallographic examinations of field replicas or of small samples removed from the in-service component. The defect size distribution is coupled with crack-growth rate data using fracture-mechanics models to compute the probable crack size as a function of time. The metallographic information is used along with material damage models and material

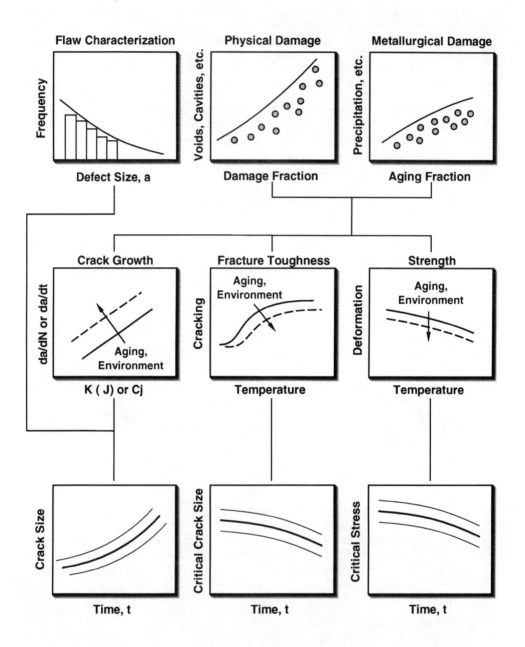

Fig. 1 Concepts for predicting the crack-growth life of structural materials

properties data to estimate crack-growth, fracture-toughness, and strength properties. As indicated in Fig. 1, these properties may be degraded by aging or exposure to an aggressive environment at high temperature. The fracture toughness and strength behavior are used to determine probable critical crack size and probable critical stress, respectively. Finally, the probable crack size and applied stress are compared with the probable critical crack size and critical stress, respectively, to predict the life of the material.

Fatigue-Crack Growth

Cyclic or fatigue-crack growth is characterized using the range and maximum value of the linear elastic stress intensity factor, ΔK and K_{max}, respectively, for conditions where the nominal cyclic stress-strain behavior is elastic. These conditions typically are satisfied for large cracks (greater than approximately 12 mm) in ductile structural materials. Thus, the cyclic crack growth rate, da/dN, is a function of ΔK and K_{max}:

$$da/dN = f(\Delta K, K_{max}) . \tag{8}$$

Equation (8) includes K_{max} as well as ΔK to account for the effect of stress ratio, R, which is the algebraic ratio of the minimum to maximum cyclic stress. The stress-ratio effect is taken into account using the Walker (1970) effective stress intensity factor, ΔK_{eff}:

$$\Delta K_{eff} = \Delta K^m (K_{max})^{1-m} , \tag{9}$$

where m is a material constant. Typical values of m are approximately 0.5. For crack growth rates above the threshold regime, Equation (8) is normally a power law as follows:

$$da/dN = D (\Delta K_{eff})^d , \tag{10}$$

where D and n are material constants.

When the nominal cyclic stress-strain behavior is inelastic, such as is typically the case for small cracks, the range of the J integral is used to characterize da/dN:

$$da/dN = f(\Delta J) . \tag{11}$$

For the a power law comparable to Equation (10), Equation (11) becomes:

$$da/dN = D (\Delta J)^d , \tag{12}$$

where D and d are material constants that can be related to D and d using the linear elastic relationship between J and K:

$$J = (1 - v^2) K^2/E \text{ for plane strain or } J = K^2/E \text{ for plane stress}, \tag{13}$$

where v is the elastic Poisson's ratio and E is the elastic modulus.

Creep-Crack Growth

Creep crack growth rate, da/dt, is characterized using the C_j parameter, which is a generalized version of the C* integral (Jaske and Begley,1978, Jaske, 1984 and 1986, Jaske and Swindeman, 1987, and Majumdar, et al., 1991). The C_j parameter is defined as follows:

$$C_j = \sigma (dv_c/dt) F , \tag{14}$$

where σ is the nominal stress, dv_c/dt is the creep displacement rate, and F is a factor that depends on the material creep behavior, crack size, and component or specimen configuration. For typical fracture-mechanics specimens, $\sigma = P/BW$, where P is the applied load, B is the thickness, and W is the width.

F factors have been developed for specimens that are commonly used in creep-crack-growth testing. For a compact-type (CT) specimen,

$$F = \{n/(n+1)\} \{2 + 0.522 (W - a)/W\}/\{1 - a/W\} , \qquad (15)$$

where n is the power-law creep exponent for the material and a is the crack length. For a center-cracked-tension (CCT) specimen,

$$F = \{(n - 1)/(n + 1)\}/\{1 - 2a/W\} . \qquad (16)$$

For an edge-notched-tension (ENT) specimen (Jaske, 1988),

$$F = 1.254 \{3.85\sqrt{n} (1 - 1/n) + \pi/n\} \{1 + 1/n\} a/g , \qquad (17)$$

where g is the effective gage length of the ENT specimen. F factors can be developed for other specimen configurations or for engineering components using the procedures discussed by Jaske (1984 and 1986) and Majumdar, et al. (1991). For example, the F-factor expression for ENT and CCT specimens have been modified and adapted to computing values of C_j for welded steam pipe (Jaske, 1990).

The C_j parameter is used to characterize da/dt in the same manner as ΔK_{eff} and ΔJ are used to characterize da/dN:

$$da/dt = f(C_j) . \qquad (18)$$

The explicit form of Equation (18) is normally a power law, as follows:

$$da/dt = \Gamma (C_j)^\gamma , \qquad (19)$$

where Γ and γ are material constants.

Creep-Fatigue Interaction

Creep-fatigue interaction effects are taken into account using a crack-tip-zone model (Jaske, 1983a). Fig. 2 presents an idealized illustration of the model. When inelastic deformation in the bulk material is small, a small process zone of diameter, r_z, exists at the crack tip. The stress levels in the process zone approach the true fracture stress, σ_f, of the material. If the fatigue crack-tip zone is smaller than a prior creep crack-tip zone or if the creep crack-tip zone is smaller than a prior fatigue crack-tip zone, an interaction effect is predicted. If the crack grows by fatigue beyond the prior creep zone or by creep beyond the prior fatigue zone, the interaction effect will no longer exist.

This interaction model is in agreement with available experimental data (Jaske, 1983a, 1983b, and 1984). A first-order estimate of the creep process zone size, r_c, is given by the following relation:

$$r_c = (1/\pi) (K_{max}/\sigma_f)^2 . \qquad (20)$$

For creep, σ_f is time-dependent and is obtained from tensile creep-rupture data. In a simi-

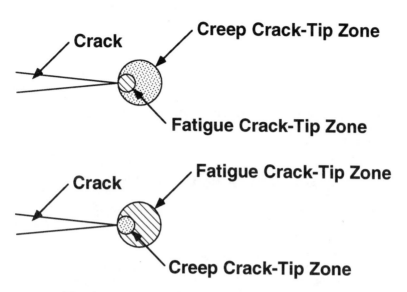

Fig. 2 Idealized illustration of crack-tip-zone model

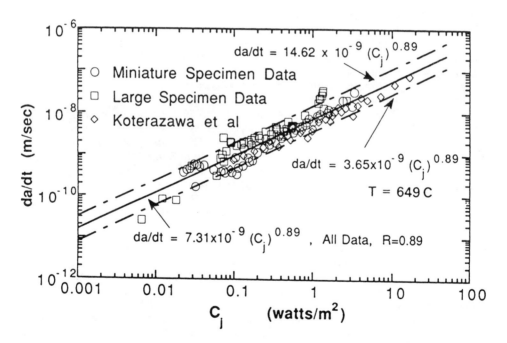

Fig. 3 Creep-crack-growth rate versus C_j for miniature and large (standard-sized) specimens of Type 316 stainless steel (from Majumdar, et al., 1991)

lar fashion, a first-order estimate of the fatigue process zone size, r_f, is given by the following relation:

$$r_f = (1/4\pi) (\Delta K/\sigma_f)^2 .$$ (21)

Comparison of Equations (20) and (21) shows that r_f is usually expected to be smaller than r_c. Thus, for most practical creep-fatigue interaction situations, fatigue crack-growth behavior is affected by prior creep crack growth. The interaction can cause either an increase or a decrease in fatigue crack growth rate. It is normally found that fatigue crack growth rate is significantly increased creep-fatigue interaction. For this reason, it is important to properly model creep-fatigue interaction in making life predictions for high-temperature structural materials.

APPLICATIONS OF CRACK GROWTH MODEL

Use of ΔK to characterize da/dN data is a widely accepted engineering practice and da/dN versus ΔK curves for structural materials are published in numerous handbooks, reports, and technical papers. Dowling (1977) has described the method for computing ΔJ for small cracks in specimens subjected to inelastic strain cycling and shown that the da/dN versus ΔJ data correlates directly with the da/dN versus ΔK data for large cracks in specimens subjected to elastic cycling. Jaske and Begley (1978) and Jaske (1983b and 1984) have reviewed the application of Dowling's method to high-temperature structural materials.

Jaske and Swindeman (1987) showed that the C_j parameter provided good characterizations of creep-crack-growth behavior for large cracks (greater than about 10 mm long) in standard-sized CT and CCT specimens of Cr-Mo-V, 2-1/4Cr-1Mo, 1-1/4Cr-1/2Mo, and 9Cr-1Mo-V-Nb steel. In a study of Type 316 stainless steel, Majumdar et al. (1991) found that the C_j parameter provided a good characterization of creep-crack-growth behavior for a wide range of specimen types and sizes, as shown in Fig. 3. The miniature specimens were of the ENT type and had a thickness of 1.47 mm, widths of 5.1 mm and 12.7 mm, and crack lengths ranging from 0.46 mm to 8.3 mm. The large specimens were standard-sized CT and CCT specimens with 25-mm to 35-mm long cracks and ENT specimens with 0.1-mm to 2-mm long cracks (Jaske, 1988). The data of Koterazawa, et al. were for tests of CCT and double-edge-notched (DEN) specimens with thicknesses ranging from 1.5 mm to 18 mm and widths ranging from 8 mm to 48 mm. Thus, the C_j parameter has been found to characterize the ductile creep-crack-growth behavior of high-temperature structural materials for a wide range of conditions.

Creep-Fatigue Crack Growth in 9Cr-1Mo-V-Nb Steel

Majumdar and Jaske (1991) evaluated the creep-fatigue crack-growth behavior of 9Cr-1Mo-V-Nb steel in air at 538 C. They tested a 1T (25.4-mm thick and 50.8-mm wide) CT specimen with 25 percent side grooves (19.1-mm net thickness) using a 24-hour loading cycle. The specimen was loaded to a maximum value of 25.6 kN and held at that value for 23 hours and 55 minutes. The load then was removed for 5 minutes to complete the loading cycle. This 24-hour loading cycle was repeated a total of 35 times during the test. The test was performed using an electrohydraulic system operated in closed-loop load control. Load was measured using a standard load cell in series with the load train, load-line displacement was measured using a high-temperature extensometer, and crack length was measured using the DC electric potential drop method.

The results of the Majumdar-Jaske test are shown in Fig. 4 where crack length is plotted as a function of time from crack initiation. The filled circular symbols represent the measured values of crack length, while the curves represent predictions made using the

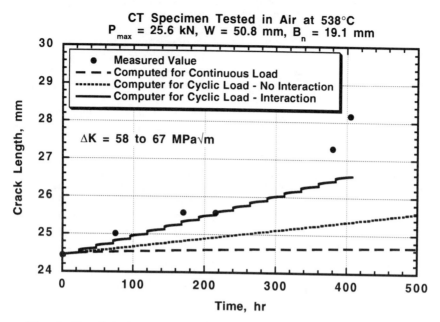

Fig. 4 Predicted versus measured creep-fatigue crack-growth behavior of 9Cr-1Mo-V-Nb steel

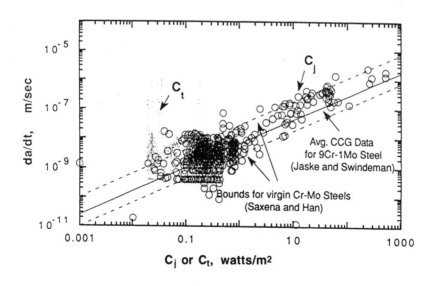

Fig. 5 Crack-growth rate of 9Cr-1Mo-V-Nb steel during 24-hour loading cycles as a function of C_j and C_t

model discussed previously. The dashed curve shows the predicted creep-crack-growth behavior for continuous loading with no cycles, while the dotted curve shows the predicted creep-crack-growth behavior for cyclic loading but no creep-fatigue interaction effect. Both of these curves significantly under-predict the actual behavior. The solid curve shows the predicted creep-crack-growth behavior for cyclic loading including the effect of creep-fatigue interaction and is in relatively good agreement with the actual data. These results illustrate that proper modeling of creep-fatigue interaction effects is required to predict such crack-growth behavior.

The crack-growth data during the hold periods at maximum load were analyzed to determine values of da/dt as a function of both the C_j parameter and the C_t parameter, as summarized in Fig. 5. C_t is the crack-tip driving force parameter developed by Saxena (1986). The methods used to compute C_t have been reviewed by Saxena, et al. (1988) and Viswanathan (1989). The circular symbols in Fig. 5 represent values of C_j, whereas the small dots represent values of C_t. The correlation of da/dt with C_t greatly under predicts the crack-tip driving force for the transient conditions during the 24-hour hold period. In contrast, C_j provides a good correlation with the steady-state creep-crack-growth data reported by Jaske and Swindeman (1987) and Saxena, et al. (1988). These results indicate that C_j characterizes creep-crack-growth behavior for transient as well as steady-state conditions.

Creep-Fatigue Crack Growth in Steam Piping

The model discussed earlier has been applied to predicting the crack-growth life of steam piping (Jaske, 1990, and Marschall, et al., 1992). Crack-growth lives have been computed for longitudinal flaws located at the internal surface and near the fusion line of seam welds in 2-1/4Cr-1Mo and 1-1/4Cr-1/2Mo steel pipes. The effects of pressure on the crack faces and local bulging of the pipe near the crack were taken into account in computing values of C_j. The influence of start/stop cycles and crack length/depth (l/d) ratios on crack-growth life were evaluated.

Fig. 6 illustrates the predicted remaining crack-growth life of a seam-welded 1-1/4Cr-1Mo steel steam pipe as a function of initial crack depths ranging from 2.54 mm to the wall thickness of 32.5 mm. The pipe had a 762-mm outside diameter (OD) and operated at 529 C under an internal pressure of 3.79 MPa. The filled circular symbols represent the predicted lives for the idealized situation of continuous operation and l/d = 10. The other symbols represent predicted lives for a cyclic operation of 2,000 hours per cycle and 4 cycles per year (8,000 total operating hours per year) and for l/d values from 2 to 20. For l/d = 10 (compare open circular symbols with filled ones), the predicted lives for cyclic operation are about half of those for continuous operation, indicating the predicted deleterious effect of cyclic operation on crack-growth life. As the l/d ratio increased, the predicted crack-growth lives decreased. Calculations of this type can be performed before in-service inspections are conducted. Then, they can be used to make disposition decisions when defect indications are found during steam-pipe inspections.

Low-Cycle Creep-Fatigue Crack Growth in Type 316 Stainless Steel

Jaske (1984) performed low-cycle creep-fatigue crack growth on both smooth and edge-notched (0.18-mm deep notch) 6.4-mm diameter specimens. The specimens were axially loaded and tested in air at 593 C and 649 C using an electrohydraulic test system operated in closed-loop strain control. Strain cycling was fully reversed with hold periods of 0 hour, 0.25 hour, and 1.0 hour at the tensile strain limit of each cycle. Cyclic crack-growth lives were determined for final crack depths ranging from 1.1 mm to 2.9 mm. For tests where creep-fatigue interaction occurred, the model discussed previously was used to predict cyclic lives. As shown in Fig. 7, the predicted lives agreed reasonably well with the measured lives.

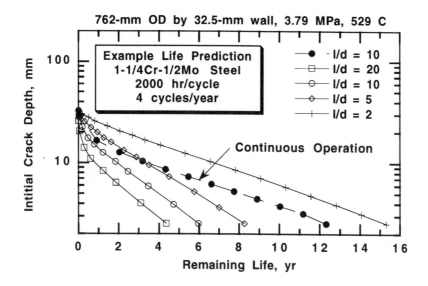

Fig. 6 Predicted remaining crack-growth life of a seam-welded 1-1/4Cr-1Mo steel steam pipe

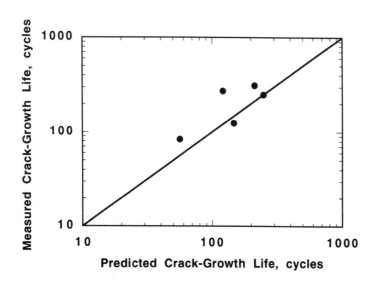

Fig. 7 Creep-fatigue crack-growth life predictions for Type 316 stainless steel (from Jaske, 1984)

SUMMARY

Methods for predicting the creep-fatigue life of high-temperature structural materials were reviewed. The traditional approaches are empirical and are based on the crack-initiation life or fracture of standard-sized, axially loaded fatigue specimens. The application of these approaches to damage-tolerant life prediction of high-temperature structural materials is limited because they do not explicitly address crack growth.

In recent years, creep-fatigue crack-growth life-prediction methods have been developed. They employ inelastic fracture mechanics (IFM) as well as linear elastic fracture mechanics (LEFM). An approach for modeling high-temperature crack growth was reviewed. Examples of its application to laboratory studies and engineering problems were discussed. Overall the approach provides reasonably accurate and useful crack-growth life predictions, including the effects of creep-fatigue interaction.

There are several areas where the crack-growth life-prediction methods need to be improved. More accurate and detailed creep constitutive equations are needed to increase the accuracy of IFM-parameter calculations. Analytical procedures for computing the C_j parameter for typical structural components also should be developed. Additional experimental work is needed to verify the life-prediction model. More experiments should be performed to develop crack-growth data for the loading histories that are encountered in engineering applications, such as long hold periods between cycles and variable amplitude and variable temperature creep-fatigue cycling. Finally, experiments should be performed on model structures or structural components to validate the analytical models for predicting high-temperature crack-growth life.

REFERENCES

Anon., 1969, "Criteria of the ASME Boiler and Pressure Vessel Code for Design by Analysis in Sections III and VIII, Division 2," ASME, New York, NY.

Basquin, O. H., 1910, "The Exponential Law of Endurance Tests," *Proceedings of ASTM*, Vol. 10, Part II, pp. 625-630.

Coffin, L. F., Jr., 1954, "A Study of the Effects of Cyclic Thermal Stresses on a Ductile Metal," *Transactions of ASME*, Vol. 76, pp. 931-950.

Coffin, L. F., Jr., 1976, "The Concept of Frequency Separation in Life Prediction for Time-Dependent Fatigue," *1976 ASME-MPC Symposium on Creep-Fatigue Interaction*, MPC-3, ASME, New York, NY, pp. 349-363.

Coffin, L. F., Jr., Manson, S. S., Carden, A. E., Severud, L. K., and Greenstreet, W. L., 1977, " Time-Dependent Fatigue of Structural Alloys - A General Assessment (1975)," ORNL-5073, Oak Ridge National Laboratory, Oak Ridge, TN.

Conway, J. B., 1969, *Stress-Rupture Parameters: Origin, Calculation and Use*, Gordon and Breach, Science Publishers, New York, NY.

Dowling, N. E., 1977, "Crack Growth During Low-Cycle Fatigue of Smooth Axial Specimens," Cyclic Stress-Strain and Plastic Deformation Aspects of Fatigue Crack Growth, ASTM STP 637, ASTM, Philadelphia, PA, pp. 97-121.

Jaske, C. E., 1983a, "A Crack-Tip-Zone Interaction Model for Creep-Fatigue Crack Growth," *Fatigue of Engineering Materials and Structures*, Vol. 6, No. 2, pp. 159-166

Jaske, C. E., 1983b, "Creep-Fatigue Crack Growth in Type 316 Stainless Steel," *Advances in Life Prediction Methods*, ASME, New York, NY.

Jaske, C. E., 1984, "Damage Accumulation by Crack Growth Under Combined Creep and Fatigue," Ph.D. Dissertation, The Ohio State University, Columbus, OH.

Jaske, C. E., 1986, "Estimation of the C* Integral for Creep-Crack-Growth Test Specimens," *The Mechanism of Fracture*, ASM International, Materials Park, OH, pp. 577-585.

Jaske, C. E., 1988, "Long-Term Creep-Crack Growth Behavior of Type 316 Stainless Steel," *Fracture Mechanics: Eighteenth Symposium*, ASTM STP 945, ASTM, Philadelphia, PA.

Jaske, C. E., 1990, "Life Assessment of Hot Reheat Pipe," *Journal of Pressure Vessel Technology*, Vol. 112, No. 1, pp. 20-27.

Jaske, C. E., and Begley, J. A., 1978, "An Approach to Assessing Creep/Fatigue Crack Growth", *Ductility and Toughness Considerations in Elevated Temperature Service*, MPC-8, ASME, New York, NY, pp. 391-409.

Jaske, C. E., and Swindeman, R. W., 1987, "Long-Term Creep and Creep-Crack-Growth Behavior of 9Cr-1Mo-V-Nb Steel", *Advances in Materials Technology for Fossil Power Plants*, ASM International, Materials Park, OH, pp. 251-258.

Koterazawa, R., and Mori, D., 1980, "Fracture Mechanics and Fractography of Creep and Fatigue Crack Propagation at Elevated Temperatures," Paper No. C 236/80, *International Conference on Engineering Aspects of Creep*, The Institution of Mechanical Engineers, London, England, pp. 219-244.

Langer, B. F., 1962, "Design of Pressure Vessels for Low-Cycle Fatigue," *Journal of Basic Engineering*, Vol. 84, No. 3, pp. 389-402.

Larson, F. R., and Miller, J., 1952, "A Time-Temperature Relationship for Rupture and Creep Stress," *Transactions of ASME*, Vol. 74, pp. 765-775.

Majumdar, B. S., and Jaske, C. E., 1991, "Creep-Fatigue Crack Growth in a 9Cr-1Mo Steel," presented at *1991 ASME Pressure Vessels & Piping Conference*, San Diego, CA.

Majumdar, B. S., Jaske, C. E., and Manahan, M. P., 1991, "Creep crack growth characterization of Type 316 stainless steel using miniature specimens," *International Journal of Fracture*, Vol. 47, pp. 127-144.

Manson, S. S., 1953, "Behavior of Materials Under Conditions of Thermal Stress," NACA TN 2933, Lewis Research Center, Cleveland, OH.

Manson, S. S., Halford, G. R., and Hirschberg, M. H., 1971, "Creep-Fatigue Analysis by Strain-Range Partitioning," *Design for Elevated Temperature Environment*, ASME, New York, NY, pp. 12-24.

Marschall, C. W., Jaske, C. E., and Majumdar, B. S., 1992, "Assessment of Seam-Welded Piping in Fossil Power Plants," EPRI Final Report TR-101835, Electric Power Research Institute, Palo Alto, CA.

Miner, M. A., 1945, "Cumulative Damage in Fatigue," *Journal of Applied Mechanics, Transactions of ASME*, Vol. 67, pp. A159-A164.

Robinson, E. L., 1952, "Effect of Temperature Variation on the Long-Time Rupture Strength of Steels," *Transactions of ASME*, Vol. 74, pp. 777-781.

Saxena, A., 1986, "Creep Crack Growth Under Non-Steady State Conditions," *Fracture Mechanics: Seventeenth Volume*, ASTM STP 905, ASTM, Philadelphia, PA, pp. 185-201.

Saxena, A., Han, J., and Banerji, K., 1988, "Creep Crack Growth Behavior in Power Plant Boiler and Steam Pipe Steels," *Journal of Pressure Vessel Technology*, Vol. 110, pp. 137-146.

Taira, S., 1962, "Lifetime of Structures Subjected to Varying Load and Temperature," *Creep in Structures*, Academic Press, New York, NY, pp. 96-124.

Viswanathan, R., 1989, *Damage Mechanisms and Life Assessment of High-Temperature Components*, ASM International, Materials Park, OH.

Walker, K., 1970, "The Effect of Stress Ratio During Crack Propagation and Fatigue for 2024-T3 and 7075-T6 Aluminum," *Effects of Environment and Complex Load History on Fatigue Life*, ASTM STP 462, ASTM, Philadelphia, PA, pp. 1-14.

AD-Vol. 36, Fatigue and Fracture of Aerospace Structural Materials
ASME 1993

LIFE PREDICTION OF A CONTINUOUS FIBER REINFORCED CERAMIC COMPOSITE UNDER CREEP CONDITIONS

T.-J. Chuang
Ceramics Division
National Institute of Standards and Technology
Gaithersburg, Maryland

ABSTRACT

This paper is concerned with development of a lifetime prediction methodology for a unidirectional fiber reinforced ceramic composite subjected to creep conditions. A continuum damage mechanics approach is adopted in which constitutive creep laws incorporating damage are constructed based on micromechanical modeling. A unit cell model is established to take advantage of the periodic feature of the material. The model which entails two elements (one representing the fiber phase and the other the matrix phase) connected in parallel is subjected to a constant stress applied in the fiber direction. From the requirements of equilibrium and compatibility, a system of simultaneous differential equations was derived for the dependent variables: stress, strain and damage as functions of time, with the initial conditions given by the elastic state of the material. The algorithm for solving this time-dependent problem was given in a flow chart from which the lifetime limited by creep can be computed. The results suggested that creep life is strongly dependent on applied stress, temperature and volume fraction of the fibers.

NOMENCLATURE

a	cavity radius	$\dot{\epsilon}_0$	reference strain rate
b	cavity half-spacing	σ_0	reference stress
d	grain size	σ_e	equivalent stress
$D_b \delta_b$	grain-boundary diffusivity	σ_f	local stress at the fiber
$D_s \delta_s$	surface diffusivity	σ_m	local stress in the matrix phase
E	Young's modulus	σ_i	principal stress (i=1,2,3)
H	crack thickness	σ_∞	applied stress (= σ_a)
K	mode-I stress intensity	Σ_s	remote tensile stress
K_G	K based on Griffith theory	Γ	materials constant
k	Boltzmann's constant	Λ	materials constant
L	length parameter at interface	Π_i	materials constant (i=1,2)
n	stress exponent	Ω	atomic volume
V_f	volume fraction of fiber	γ_b	grain-boundary free energy
V_{min}	minimum crack velocity	γ_s	surface free energy
T	temperature	ν	Poisson's ratio
t	exposure time	ω	damage parameter
t_r	rupture time or lifetime		

INTRODUCTION

In recent years, advanced ceramics such as continuous fiber reinforced ceramic matrix composites (CFCC) have attracted considerable attention due to certain advantages relative to conventional materials (e.g., super-alloys) in structural applications at elevated temperatures. Those advantages include increased strength, enhanced toughness, high creep and corrosion resistance in demanding service environments. The load history of a typical structural member or component, regardless of its application, normally includes a monotonically increasing load regime followed by a long period of sustained load, plus sporadic, unsteady, or cyclic thermal-mechanical load excursions. Therefore, if CFCCs are to be successfully and confidently used in high temperature, load-bearing applications they must survive the initial increasing load regime, *as well as* remain reliable during the steady-state portion of the design life. Unfortunately, the lack of a design methodology for assessing service life dominated by creep rupture is one of the major hurdles preventing widespread application of these emerging materials. There are several factors responsible for the present deficiency: (1) since fabrication of CFCCs is still in the developmental stage, material parameters are either unknown or constantly being improved; (2) collection of long-term test data under sustained loading conditions is time consuming, and the data collected may be obsolete and irrelevant since the material evolves into a new system due to fabrication improvements; and last, but not least, (3) a lack of theories which allow extrapolation of short-term laboratory data to long-term service conditions. The last item is particularly critical for long-term applications since theoretical models are required (in addition to the experimental data necessary for verification purposes) for any material system used in the fabrication of structural components.

This paper focuses on the last issue and presents a methodology from which a design engineer can obtain an estimate for the creep rupture lifetime. Since the material under consideration is aimed at applications where loads are of long duration, the design engineer is interested in the material's creep response at low or intermediate levels of sustained loading. It is now well recognized that at the lower end of the applied stress spectrum creep damage usually appears in the form of cavities during the majority of service life. The cavities are distributed in a heterogenous fashion along grain-boundary facets that have orientations coincident with the directions normal to the maximum principal tensile stress. Macroscopic crack growth (i.e., subcritical crack growth) occurs only towards the end of the creep life as a result of microcrack linkage that emerges from the coalescence of cavities growing at the grain boundaries. Under this scenario macroscopic crack growth will be insignificant during a major portion of the life of a structural component. In addition, it seems suitable to adopt continuum damage mechanics concepts such that a damage parameter is incorporated in the constitutive laws in order that the effects of microdefects are represented at the macroscopic level. It should be recognized, however, that if damage is dominated by the single crack growth mode, this approach is not applicable.

The present paper is organized in the following manner. First, constitutive laws are presented for a variety of creep damage cases observed in advanced ceramics. Those constitutive creep laws (including damage) are then applied to the unidirectional CFCC subjected to uniaxial tensile creep in the fiber direction. For the sake of brevity, discussion is limited to only uniaxial formulations. Due to the periodic structure of CFCC, a unit cell model is adequate to characterize the creep behavior. By modeling the fiber and matrix as separate phases connected in parallel, a system of (coupled) ordinary differential equations of first order for stress, strain and damage is obtained. The analysis assumes that the constituents are subject to uniaxial states of stress (i.e, a one dimensional analysis). However, using this technique it is possible to acquire solutions for stress, strain and damage states as functions of time. The approach can be used as a screening technique where a creep rupture lifetime can be estimated from the time at which damage in the matrix or fiber reaches a critical value.

74

CONSTITUTIVE CREEP LAWS

In classical continuum mechanics, the constitutive laws usually do not include damage as an active parameter so that lifetime limited by damage cannot be estimated from the discipline. To predict creep life, the concepts of continuum damage mechanics incorporating damage as a major variable must be adopted. However, there is no unique way of defining damage. In the case of creep deformation, damage could be defined as a crack density which is equal to the number of cracks per unit volume times the effective cracked volume (e.g., a^3, see Budiansky and O'Connell [1], Rodin and Parks [2], Duva and Huchinson [3], Hasselman, et al. [4]). Alternatively, damage could be defined as the cavity volume fraction along the grain-boundary of a volume element (Cocks and Leckie [5]) or the ratio of cavitated area over total grain-boundary area (Cocks and Ashby [6]). Different authors have pointed out that for a general anisotropic material subject to multi-axial stress states, damage must have a tensorial character. Damage can be quantified either by a fourth or eighth order tensor (see Ju [7], or Chow and Wang [8]). However, for macroscopically isotropic and homogeneous solids subject to uniaxial tensile creep, it is sufficient to characterize damage in terms of fraction of total cavitated grain-boundary area (Cocks and Ashby [6], Cocks and Leckie [5]). Utilizing this method as a measure of damage, it is possible to develop several constitutive laws for uniaxial creep while simultaneously incorporating continuum damage principles. These constitutive laws can be derived from micro-mechanistic models that have found application to ceramic materials. These micro-mechanistic models include power-law creep due to grain-boundary sliding, lenticular cavity growth in rigid grains, and crack-like cavity growth in rigid as well as elastic grains. Other possible kinetic laws are outlined by Cocks and Ashby [6], where kinetic laws that address dislocational creep, transgranular hole growth, and grain-boundary reaction with precipitates are presented. These mechanisms are more relevant to creep in metal alloys and will not be discussed here.

Phenomenological Power-law Creep

In the absence of damage, ceramic composites with polycrystalline fibers and matrix can deform by power-law creep. This assumes that the material is fully dense and subject to very low stress levels. Here the steady-state creep rate takes the form:

$$\dot{\epsilon} = \dot{\epsilon}_0 \left(\frac{\sigma_e}{\sigma_0} \right)^n \qquad (1)$$

where σ_0 is the reference stress, and the reference strain rate $\dot{\epsilon}_0$, and the stress exponent n are materials constants. Here σ_e is an equivalent stress defined by $\sigma_e = [(3/2)(S_{ij}S_{ij})]^{1/2}$ where S_{ij} is the deviatoric stress tensor. Alternatively the equivalent stress can be expressed in terms of the three Cauchy principal stresses $(\sigma_1, \sigma_2, \sigma_3)$ as $\sigma_e = \{(1/2)[(\sigma_1 - \sigma_2)^2 + (\sigma_1 - \sigma_3)^2 + (\sigma_2 - \sigma_3)^2]\}^{1/2}$. Notice that for simple uniaxial tension $\sigma_1 = \sigma$ and $\sigma_2 = \sigma_3 = 0$, thus $\sigma_e = \sigma$.

For this particular classical formulation, power-law creep due to diffusion (n=1) occurs in the matrix by lattice diffusion (Nabarro [9], Herring [10]), diffusion along grain boundaries (Coble [11]), or through solution-reprecipitation (Chen [12]). Alternatively, diffusion that is accommodated by grain boundary sliding (n=2) results in grain rotation (Ashby and Verrall [13]) or the emergence of new grains (Gifkins [14]). However, literature reviews that focus on the subject of power-law creep in structural ceramics indicate that n ranges from 1 to 8, and in some instances n is reported in excess of 10. It is important to note that most values are consistently higher than 1 or 2, i.e., higher n values than predicted by conventional theories (Cannon and Langdon [15]). This discrepancy points to the need for taking creep damage into account in the formulation of constitutive creep laws. In the following sections, three prominent creep damage mechanisms observed in advanced ceramics are discussed, and constitutive equations for creep and damage rates are derived. These rates are expressed in terms of stress, a damage parameter, temperature and other relevant material properties.

Growth of Lenticular Cavities in Rigid Grains

Consider a lenticular cavity with radius a located at a grain boundary with an area A. The area fraction $(\pi a^2/A)$ is subjected to an applied far field uniaxial stress σ_∞ in a direction normal to A. For growth of a lenticular cavity in a rigid grain, the following conditions must be met: (1) surface diffusivity dominates (i.e., grain-boundary diffusivity is minimal) such that $\Delta \equiv (D_s\delta_s/D_b\delta_b) \gg 1.0$; (2) the applied stress is low in comparison to the capillary (or sintering) stress; and (3) the focus is on the early stage of creep (Chuang, et al. [16]). Furthermore, in order to have the matrix behave rigidly, cavities must be closely spaced, i.e., cavity spacing must be small compared to cavity size a so that material diffused from the cavity surfaces can be distributed *uniformly* along the grain boundary. Fig. 1 is a transmission electron microscope (TEM) photograph taken from a post crept silicon nitride specimen subjected to a tensile stress of 125 MPa at 1350°C for 125 hours [17]. As indicated in the photograph, lenticular cavities were formed at a grain-boundary facet.

Growth rates for lenticular cavities have been solved for by several authors (see Hull and Rimmer [18]; Raj and Ashby [19]; or Speight and Harris [20]). By approximating the cavity shape as spherical and neglecting surface tension effect, the evolution of ω (defined as a^2/b^2 where $2b$ is the cavity spacing) is given by

$$\dot{\omega} = \frac{\Gamma\dot{\epsilon}_0}{\sqrt{\omega}\,\ell n(1/\omega)}\left(\frac{\sigma_\infty}{\sigma_0}\right) \tag{2}$$

where $\Gamma = 2D_b\delta_b\Omega\sigma_0/kTb^3\dot{\epsilon}_0$ is a material constant depending on atomic volume Ω, geometry, temperature T, and grain-boundary diffusivity, i.e., $D_b\delta_b$. Here damage is represented as a scalar quantity, and damage evolution follows a linear dependence on stress. However, the damage rate depends on the current damage state in a nonlinear fashion. A closed form solution for creep rupture life is obtained by integrating Eq. 2, provided that the initial damage state is known and the stress is held constant throughout the load history (Cocks and Ashby [6]).

The creep strain rate induced by lenticular cavity growth assumes that the grain is non-deforming. In essence the strain rate is the result of a mechanism that uniformly distributes material at the grain-boundary ligament in a jacking fashion. The resulting strain rate is given by the expression

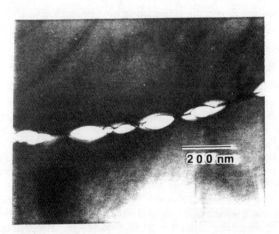

Figure 1 TEM photograph showing lenticular creep cavities at a grain boundary.(Ref. [17])

$$\dot{\varepsilon} = \frac{2\Gamma\dot{\varepsilon}_0}{\ell n(1/\omega)}\left(\frac{b}{d}\right)\left(\frac{\sigma_\infty}{\sigma_0}\right) \tag{3}$$

As was the case for the damage rate, the inelastic strain rate is linearly dependent on stress, and increases in a nonlinear fashion with increasing ω. It should be noted that Eq 3 only represents that portion of the strain rate contributed by cavity growth. The total strain rate is the summation of contributions from all active deformation mechanisms (including elastic deformations).

Growth of Crack-like Cavities in Rigid Grains

When surface diffusion is slow in comparison to grain-boundary diffusion, or when the applied stress is high in comparison to the sintering stress, or during later stages of creep life, cavities grow into crack-like shapes (Chuang, et al. [16]). The shape of the growing crack at steady state conditions is controlled by surface diffusion. The solution obtained by Chuang and Rice [21] indicates a constant crack thickness H is developed, i.e.,

$$H = 2\sqrt{2}\sqrt{1-\frac{\gamma_b}{2\gamma_s}}\left(\frac{D_s\delta_s\gamma_s\Omega}{\dot{a}kT}\right)^{1/3} \tag{4}$$

Note that a thinning crack will develop with high crack velocity or applied stress. This latter result is interesting since it contradicts the behavior of an elastic crack opening displacement where a larger opening occurs with higher stresses.

The growth of a crack-like cavity in a rigid grain has been considered by Chuang, et al. [16]. The damage rate in the lower stress regime can be cast in the following form:

$$\dot{\omega} = \frac{\Pi_1\sqrt{\omega}\dot{\varepsilon}_0}{(1-\omega)^3}\left(\frac{\sigma_\infty}{\sigma_0}\right)^3 \tag{5}$$

where $\Pi_1 = D_s\delta_s\Omega\sigma_0^3/(\sqrt{2}kTb\varepsilon_0\gamma_s^2)$. The creep strain rate due to jacking (i.e., the removal of material from the crack surfaces via surface diffusion and the resulting uniform deposition along the grain boundary via grain-boundary diffusion) takes the following form (see Chuang, et al. [16], Cocks and Ashby [6], or Cocks and Leckie [5]):

$$\dot{\varepsilon} = \frac{\Pi_2\sqrt{\omega}\dot{\varepsilon}_0}{(1-\omega)^3}\left(\frac{\sigma_\infty}{\sigma_0}\right)^2 \tag{6}$$

where $\Pi_2 = 4\Pi_1\gamma_s/d\sigma_0$.

Growth of Crack-like Cavities in Elastic Grains

In the early stages of creep when the damage is minimal, crack-like cavities growing along grain boundaries will not interact since neighboring cracks are remotely located due to large cavity spacings. Here the grains must deform elastically to accommodate material deposited primarily in the near-tip zone. A diffusive crack growth model has been presented by Chuang [22] to address this situation. In this model plane-strain conditions are assumed so that measures are based on a unit thickness. This is the only self-consistent geometry possible, i.e., the only geometry where the crack front will remain straight during growth. Once growth commences for penny shaped cracks, the axi-symmetric properties of the crack are destroyed due to crack interactions. A typical crack-like cavity is shown in Fig. 2 where a TEM photograph depicts a portion of an alumina bend bar near the tensile edge [23]. The alumina bar was subjected to creep deformation under four-point bend load conditions at $1720°C$. The crack length is approximately 8 micrometers, and the thickness is approximately 125 nanometers. The direction of the applied stress is vertical in this microphotograph.

Solving the integro-differential equation for the unknown stress distribution at the grain boundary, the crack growth rate $\dot{a}$ can be expressed in the following form (Chuang [22]):

$$\dot{a} = V_{min}\left[0.59\frac{K}{K_G} + \sqrt{0.35(\frac{K}{K_G})^2 - 1}\right]^{12} \tag{7}$$

where $K \equiv \sigma_\infty\sqrt{b}[\tan(\omega/2)]^{1/2}$ and K_G and V_{min} are material constants defined as

$$K_G = \sqrt{\frac{E(2\gamma_s - \gamma_b)}{(1-\nu^2)}} \tag{8}$$

and

$$V_{min} = 8.13\frac{D_s^4\Omega^{7/3}}{kT\gamma_s^2}\left[\frac{E}{(1-\nu^2)D_b\delta_b}\right]^3 \tag{9}$$

The damage parameter ω is defined as the area fraction of cavitated area over total grain-boundary facet ($\omega = a/b$) where b is half the center-to-center crack spacing. Thus $\dot{\omega} = \dot{a}/b$, and the damage rate is

$$\dot{\omega}(\omega,\sigma_\infty) = \left(\frac{V_{min}}{b}\right)\left[0.59\frac{K}{K_G} + \sqrt{0.35\left(\frac{K}{K_G}\right)^2 - 1}\right]^{12} \tag{10}$$

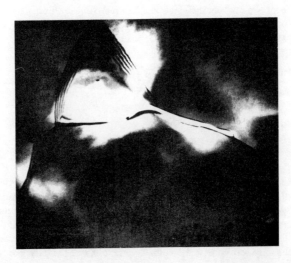

Figure 2   TEM microphoto of a crack-like creep cavity. (After Chuang and Tighe [23])

After the material is deposited non-uniformly in the vicinity of the grain boundary near the crack-tip, the displacement opening at the grain boundary widens, and the inelastic strain rate is approximated by $\dot{\epsilon} = <\delta>/d$ where $<\delta>$ is the average opening displacement. Thus $\delta = \dot{a}(\delta/x) = (\dot{a}H/L)(-\partial\delta/\partial\hat{X})$, where L is a length parameter defined as $L = (ED_b\delta_b\Omega/\dot{a}kT)^{1/2}$. Examination of the solution of $<\delta>$ given by Chuang [22] indicates that $(-\partial\delta/\partial\hat{x})=0.5$. Accordingly, the strain rate takes the form $\dot{\epsilon} = \dot{a}H/(2dL) = \Lambda\dot{a}^{7/6}/d$ or

$$\dot{\epsilon}(\omega,\sigma_\infty) = \left(\frac{\Lambda}{d}\right)\left(\frac{V_{min}}{b}\right)^{7/6}\left[0.59\frac{K}{K_G} + \sqrt{0.35\left(\frac{K}{K_G}\right)^2 - 1}\right]^{14} \tag{11}$$

where

$$\Lambda = \left(\frac{\gamma_s^2\Delta^2kT}{E^3D_b\delta_b\Omega}\right)^{1/6} \tag{12}$$

In this section constitutive creep laws have been derived that are based on a number of well-accepted cavity growth models. In addition the laws also incorporate a damage parameter. In the next section we will apply these constitutive laws to estimate the creep life of a CFCC using a simple unit cell model.

A UNIT CELL MODEL

Now consider a CFCC with a volume fraction of fibers V_f, that is subjected to a constant remote tensile stress Σ_a such that the material creeps. A cross-section of the material under consideration is depicted in Fig. 3(a) where the direction of the applied load is parallel to the fiber orientation. Fig. 3(b) is a schematic of the parallel connection model representing the behavior of a unit cell, which is indicated by dashed lines in Fig. 3(a). The model consists of two elements, one representing matrix material, and the other representing the fiber element. Both are connected in parallel and are subjected to far-field (remote) constant stress that gives rise to a creep strain rate. A similar exercise with metal matrix composites (MMC) has been considered by Goto and McLean [24]. A basic assumption of the model is that the fiber is perfectly bonded to the matrix phase so that no displacement discontinuities occur across the fiber/matrix interface. An analysis by Hill, et al. [25] made on a SiC/RSBN composite indicated that this condition can be fulfilled by a finite element model representing a similar unit cell approach.

As a result of compatibility requirements, the strain rate within fiber and matrix must be equal to the remote creep strain rate. However, the stress states within the two elements will change continuously with time, although the macroscopic (or homogenized) stress remains constant. In continuum mechanics terms, this modeling approach is analogous to the idea of "constrained" cavitation proposed by Dyson [26] and Rice [27]. Denoting the time-dependent stress in the fiber and matrix as σ_f and σ_m respectively, equilibrium conditions require

$$\Sigma_a = \sigma_f V_f + (1-V_f)\sigma_m \tag{13}$$

Since Σ_a is constant, we have $\dot{\sigma}_f = -(1 - V_f)(\dot{\sigma}_m/V_f)$ or

79

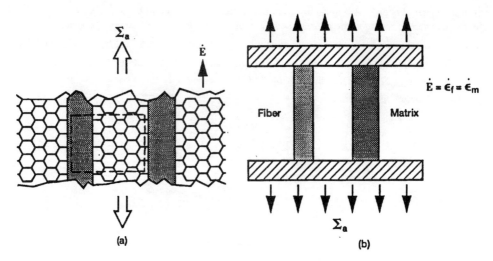

Figure 3 Modeling of creep in CFCC: (a) schematic of the material; (b) unit cell model.

$$\sigma_f(t) = \epsilon_0\left[E_f + \frac{E_m(1-V_f)}{V_f}\right] - \left[\frac{1-V_f}{V_f}\sigma_m(t)\right] \tag{14}$$

where $\epsilon_0 = E(0) = \Sigma_a/[E_m(1-V_f)+E_fV_f]$ is the initial elastic strain in either matrix or fiber, and the initial stresses in the fiber and matrix are $\sigma_f = E_f\epsilon_0$ and $\sigma_m = E_m\epsilon_0$, respectively. The total strain includes an elastic portion as well as inelastic strains induced by deformation mechanisms discussed previously. In the case of ceramic fibers, results from creep tests on free fiber bundle indicate that fibers usually suffer power-law creep due to grain-boundary sliding in addition to crack damage (see DiCarlo [28], and Tressler [29]). The reader is cautioned that power-law creep in an unconstrained fiber may differ from a fiber that is constrained by the matrix within a CFCC. Fibers within a CFCC are usually constrained such that sliding may be difficult to activate. Accordingly, the rate of fiber strain at any time can be expressed as

$$\dot{\epsilon}_f = \frac{\dot{\sigma}_f}{E_f} + A_f\sigma_f^n + F_f(\sigma_f, \omega_f) \tag{15}$$

where F_f is the strain rate obtained from either Eq. 3, Eq. 6 or Eq. 11, depending on the material and the active damage mechanism. Similarly, the total strain rate in the matrix is

$$\dot{\epsilon}_m = \frac{\dot{\sigma}_m}{E_m} + A_m\sigma_m^N + F_m(\sigma_m, \omega_m) \tag{16}$$

where, again, F_m is the strain obtained from either Eq. 3, Eq. 6 or Eq. 11. Now with $E = \epsilon_f = \epsilon_m$, from compatibility requirements the following expression

$$\dot{\sigma}_m = \left[\frac{E_m E_f V_f}{E_m + V_f(E_f - E_m)}\right] F(\sigma_m) \tag{17}$$

80

is obtained by equating Eq. 15 and Eq. 16 and solving for $\dot{\sigma}_m$, subjected to the initial condition $\sigma_m(0) = E_m \, \varepsilon_0$. Hence, a numerical algorithm can be developed to solve for the matrix and fiber stresses, the total strain and damage within both the fiber and matrix at every time step. The algorithm is as follows:

(1)   at t=0 $\sigma_m(0)$, $\sigma_f(0)$, $\omega_m(0)$ and $\omega_f(0)$ are known;

(2)   at t= $t_0+\Delta t$, compute $\sigma_m(t)$ from Eq. 17;

(3)   knowing $\sigma_m(t)$, $\sigma_f(t)$ is calculated from Eq. 14 and E(t) is calculated from Eq. 16;

(4)   similarly, ω can be calculated from Eq. 2, Eq. 5 or Eq. 10;

(5)   steps (2) and (3) are repeated to obtain the stresses, strain and damage values until both ω_f and ω_m reach a critical value;

(6)   the time at which ω_f or ω_m reach a critical value becomes the creep life (t_r) of the material and computations cease.

A flow chart depicted this numerical procedure is given in Fig. 4. Based on this flow chart, a FORTRAN computer program can be written to solve the problem. Fig. 5 depicts a generic plot of the evolution of stress within the fiber and matrix, the local strain, and damage in a typical unit cell. Here it is assumed that the fibers are stiffer ($E_f > E_m$)and creep at a slower rate than the ceramic matrix. Thus the fibers stresses are higher than the ceramic matrix, and damage accumulates more in the fiber resulting in earlier fracture than the matrix. In this way, the effect on the rupture life of material properties (e.g., volume fraction of fibers), and other parameters such as stress and temperature can be studied.

It should be reiterated that this unit cell model assumes a perfect bonding between the two constituents so that the whole body is simply a building block of these unit cells. In the case where local slips between fiber and matrix do occur as indicated by the general phenomena of fiber pull-out in many conventional composites, this model is inapplicable and a finite element approach using

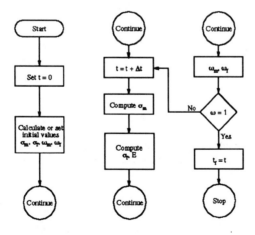

Figure 4 Flow chart for calculating the creep life

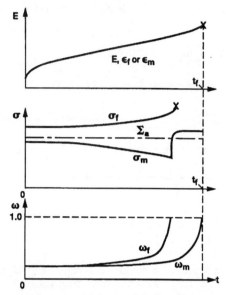

Figure 5 Schematic plots of strain, stress and damage as functions of time.

some type of interface or interphase elements may have to be implemented. Unlike the unit cell where creep strain is independent of location, it becomes a field parameter and will vary throughout the body. We will leave this for future study, together with the problems of reliability in a CFCC of 2D fiber architecture.

CONCLUDING REMARKS

A methodology to estimate creep service life based on continuum mechanics concepts has been presented. A numerical algorithm is presented to solve the initial value problem in which stress, strain and damage states can be computed in a time-step fashion, if the constitutive creep laws which dictate the materials behavior are given. A number of creep damage mechanisms that are most likely active in advanced ceramics are examined, and their corresponding constitutive laws, which incorporate damage, are established. This methodology is applied to estimate the creep life of CFCC (a promising aerospace material). We found that creep life is strongly dependent on the applied stress, temperature and volume fraction of the fiber.

ACKNOWLEDGEMENT

This work was partially supported by HITEMP program, NASA Lewis Research Center (Project Monitor is Dr. John P. Gyekenyesi) under Interagency Agreement C-82000-R between NASA and NIST.

REFERENCES

1.   Budiansky, B. and O'Connell, R.J., "Elastic Moduli of a Cracked Sold," International Journal of Solids and Structures, Vol. 12, January 1976, pp 81-97.

2. Rodin, G.J. and Parks, D.M., "Constitutive Models of a Power-law Matrix Containing Aligned Penny-Shaped Cracks," Mechanics of Materials, Vol. 5, No. 3, September, 1986, pp 221-228.

3. Duva, J.M. and Hutchinson, J.W., "Constitutive Potentials for Dilutely Voided Nonlinear Materials," Mechanics of Materials, Vol. 3, 1984, pp 41-54.

4. Hasselman, D.P.H., Donaldson, K.Y. and Venkateswaran, A., "Observations on the Role of Cracks in the Non-linear Deformation and Fracture Behavior of Polycrystalline Ceramics," in Fracture Mechanics of Ceramics, Vol. 10, Eds. R.C. Bradt, D.P.H. Hasselman, D. Munz, M. Sakai and V. Ya. Shevchenko, Plenum Press, New York, 1992, pp 493-507.

5. Cocks, A.C.F. and Leckie, F.A., "Creep Constitutive Equations for Damaged Materials," Advances in Applied Mechanics, Eds. J.W. Hutchinson and T.Y. Wu, Academic Press, Vol. 25, 1987, pp 239-294.

6. Cocks, A.C.F. and Ashby, M.F., "On Creep Fracture by Void Growth," Progress in Materials Science, Vol. 27, 1982, pp 189-244.

7. Ju, J.W., "A Micromechanical Damage Model for Uniaxially Reinforced Composites Weakened by Interfacial Arc Microcracks," Journal of Applied Mechanics, Vol. 58, December 1991, pp 923-930.

8. Chow, C.L. and Wang, J., "An Anisotropic Theory of Continuum Damage Mechanics for Ductile Fracture," Engineering Fracture Mechanics, Vol. 27, May 1987, pp 547-558.

9. Nabarro, F.R.N., "Steady-State Diffusional Creep," Philosophical Magazine, Vol. 16, 1967, pp 231-237.

10. Herring, C., "Diffusional Viscosity of a Polycrystalline Solid," Journal of Applied Physics, Vol. 21, 1950, pp. 437-445.

11. Coble, R.L.,"A Model for Boundary Diffusion Controlled Creep in Polycrystalline Materials," Journal of Applied Physics, Vol. 34, June 1963, pp 1679-1682.

12. Chen, C.F., "Creep Behavior of Sialon and Siliconized Silicon Carbide Ceramics," Ph.D. Dissertation, University of Michigan, Ann Arbor, MI, 1987.

13. Ashby, M.F. and Verrall, R.A., "Diffusion Accommodated Flow and Superplasticity," Acta Metallurgica, Vol. 21, January 1973, pp 149-163.

14. Gifkins, R.C., "Grain Rearrangements during Superplastic Deformation," Metallurgical Transactions, Vol. 13, November 1978, pp 1926-1936.

15. Cannon, R.W. and Langdon, T.G., "Review : Creep of Ceramics, Part I: Mechanical Characteristics," Journal of Materials Science, Vol. 18, January 1983,pp 1-50.

16. Chuang, T.-J., Kagawa, K.I., Rice J.R., and Sills, L.B., "Overview No.2: Non-equilibrium Models for Diffusive Cavitation of Grain Interfaces," Acta Metallurgica, Vol. 27, February 1979, pp 265-284.

17. Hockey, B. J., Wiederhorn, S.M., Liu, W., Baldoni, J. G. and Bujin, S.-T.: Tensile Creep of Whisker-Reinforced Silicon Nitride, Journal of Materials Science, Vol. 26, No. 1, 1991, pp 3931-40.

18. Hull, D. and Rimmer,D.E., "The Growth of Grain-Boundary Voids under Stress," Philosophical Magazine, Vol.4 [42] 1959, pp 673-687.

19. Raj, R. and Ashby, M.F., "Intergranular Fracture at Elevated Temperature," Acta Metallurgica, Vol. 23, June 1975, pp 653-666.

20. Speight, M.V. and Harris,J.E., "Kinetics of Stress-Induced Growth of Grain-Boundary Voids," Metal Science Journal, Vol. 1 [1] 1967, pp 83-85.

21. Chuang T.-J. and Rice, J.R.,"The Shape of intergranular creep Cracks Growing by Surface Diffusion," Acta Metallurgica, Vol. 21, No. 12, December 1973, pp 1625-1628.

22. Chuang, T.-J., "A Diffusive Crack Growth Model for Creep Fracture," Journal of the American Ceramic Society, Vol. 65, No. 2, February 1982, pp 93-103.

23. Chuang T.-J. and Tighe, N.J., "Diffusional Crack Growth in Alumina," in Proceedings of the 3rd International Conference on Fundamentals of Fracture, pp 129-132, Ed. P. Neumann, Max-Plank-Institut für Eisenforschung GmbH, Düsseldorf, Germany, June, 1989.

24. Goto, S. and McLean, M., "Role of Interfaces in Creep of Fibre-reinforced Metal-Matrix Composites - I. Continuous Fibres," Acta Metallurgica et Materialia, Vol. 39, No. 2, February, 1991, pp 153-164.

25. Hill, B.B., Hahn, H.T., Bakis, C.E. and Duffy, S.F., "Micromechanics-Based Constitutive Relations for Creep of SiC/RSBN Composites," in HITEMP Review 1991, NASA CP-10082, 1991, pp. 58-1 through 58-13.

26. Dyson, B. F., "Constrained Creep Cavitation," Metal Science, Vol. 10, 1976, pp 349-353.

27. Rice, J.R., "Constraints on the Diffusive Cavitation of Isolated Grain Boundary Facets in Creeping Polycrystals," Acta Metallurgica, Vol. 29, 1981, pp 675-681.

28. DiCarlo, J. A. "Creep of Chemically Vapor Deposited SiC Fibers," Journal of Materials Science, Vol. 21, January 1986, pp 217-224.

29. Tressler, R.E. and Pysher, D.J., "Mechanical Behavior of High Strength Ceramic Fibers at High Temperatures," in Proceedings of 7th CIMTEC World Ceramic Congress, Ed. P. Vincenzini, Elsevier Science Publishers, Amsterdam, Netherlands, 1991.

AD-Vol. 36, Fatigue and Fracture of Aerospace Structural Materials
ASME 1993

AN EQUIVALENT DAMAGE CONCEPT
FOR FATIGUE LIFE ANALYSIS

Arvind Nagar
Flight Dynamics Directorate
Air Force Wright Laboratory
Wright-Patterson AFB, Ohio

ABSTRACT

Recent developments in fracture mechanics basic research relating to life prediction methods of aerospace structures are discussed. The effects of structural crack geometry and flight induced loads on life are two of the most important considerations when extending life prediction techniques developed at the material scale to applications in the development, design and operational maintenance of aerospace structures. The variable-load interaction models for fatigue crack growth are reviewed. Some recent analytic and numerical solutions of stress at holes and notches of aerospace interest are discussed. An equivalent damage concept to predict life of structural elements subjected to variable loads is introduced.

I. INTRODUCTION

This paper presents life analysis and life prediction techniques used in the development and design of aerospace structures. Other papers in this ASME volume discuss these techniques for applications to other structures as presented at the life prediction session of the fatigue and fracture symposium. The methods for structural life analysis in the aerospace environment require structural material response characterization, organization of flight loads to formats compatible to analysis and testing, determination of fracture damage parameters from applied loads, incorporation of the effects of variable loads on damage, development of damage progression models, damage accumulation schemes and a criteria for failure. The final life analysis is verified by performing tests simulating usage and the structural conditions. For conditions of self similar cracking satisfying similitude criteria, the current methods have been found to be useful in the design and maintenance of structures. However, due to variability in material properties, manufacturing quality and the prediction of operating conditions, the monitoring of aircraft structures and establishment of conditions for retirement or continued operation requires updating the current methods using the latest developments in fracture and damage technologies with a view to improving their accuracy of prediction.

The airframe is required to resist fatigue cracking, corrosion, thermal degradation, delamination and wear during service to assure economic life (Figure 1). The safety of flight structure requirements include flaw existence assumptions and residual strength and damage growth limits. The crack growth and residual strength analyses are required to assure damage tolerant structures. It is required that the analyses assume the presence of flaws placed in the most unfavorable location and orientation with respect to the applied stresses and the material properties. The crack growth analysis is required to predict the growth behavior of these flaws in the chemical, thermal, and sustained and cyclic stress environments to which that portion of the component shall be subjected to in service. The spectrum interaction effects, such as variable loading and environment are to be accounted for.

The life analysis methods to satisfy the above requirements are discussed in this paper. An equivalent damage concept is proposed to predict life of structures subjected to variable loads such as those illustrated in Fig. 2 by transforming variable loads to equivalent constant amplitude cyclic loads. The load block in Figure 2 includes the overloads, the underloads, loads with hold times, interrupted loads and constant amplitude cyclic loads. The transformation of such load variations is based on equating the damage caused by variations in loads and associated spectrum and environmental interaction effects as shown in the Figure.

II. CRACK GROWTH ANALYSIS

Crack growth in aerospace structures is affected by the structural crack geometry and the encountered loads during flight. The crack growth rate is shown to depend on the crack-tip stress intensity factor range, stress ratio, thickness, temperature, frequency and the air environment.

$$\frac{da}{dN} = f(\Delta K, \ R, \ t, \ T, \ f, \ env.) \tag{1}$$

It is to be noted that the width of a specimen can also affect crack growth apart from its indirect effect through the stress intensity factor. During fatigue crack growth tests, three regions of crack growth rate behavior with the cyclic stress intensity factor are observed. The continuum

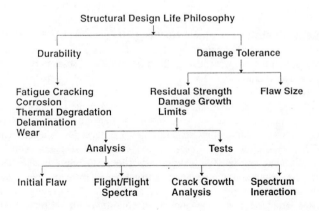

Figure 1. Durability and Damage Tolerance Components of Structural Design Life Philosophy.

fracture mechanics region is distinguished by a linear log-log relation. The third region is too close to instability and contributes little to crack life. The low crack growth region carries the most service life in some areas but is relatively less understood because of the micro-level interactions. For this reason, damage tolerance analysis requires assuming an initial flaw size for which the continuum approach is usually valid. Some of the required tasks for crack growth analysis are shown in Fig. 3. The baseline crack growth data is obtained on small material coupons. The stress intensity factor solutions for the crack geometry are obtained. The requirements for analysis verification testing is one of the most important requirements for damage tolerance. A crack in a structure is likely to be surrounded by stiffeners, holes, fasteners, and treated surfaces, and often have different shapes and planar aspect ratios (i.e., quarter circular, quarter elliptical, semi-circular and semi-elliptical shapes). Such cracks may nucleate anywhere in the material volume. The stress intensity factors are normally adjusted to account for such effects. There are several computational solutions for the stress intensity factor

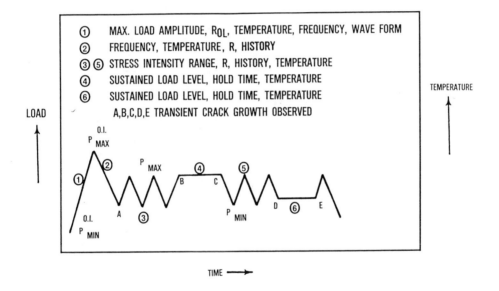

Figure 2. Spectrum Interaction Effects Under Mechanical and Thermal Loads

solutions for such crack shapes. In aerospace structures, a typical crack geometry is a corner crack at a hole. A finite element based stress intensity factor solution for such a crack was obtained by Newman and Raju[1] and is discussed by Harmon and Saff[5]. The results are presented in Appendix-1. The effects of thickness, width, crack planar aspect ratio and the hole size are incorporated. The stress intensity factor distributions at the front of such cracks is non-uniform and thus each point at the front would tend to grow at a different rate.

[x] denotes Reference x

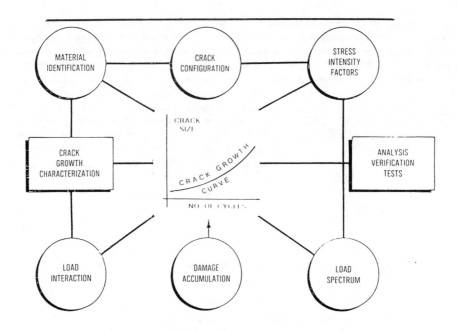

Figure 3. Crack Growth Analysis Methodology for Damage Tolerance

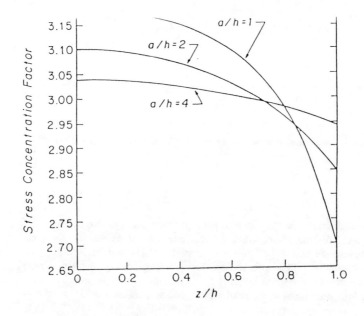

Fig. 4. Effect of Structural Geometry on Stresses at a Hole in a
Three-Dimensional Plate

Due to its importance, the structural hole has been a subject of a number of current studies. A three-dimensional analytic solution of the stresses at a hole was recently obtained by Folias and Wanget[2]. The results are presented in Figure 4 and show that the maximum stresses depend on the hole size. The stresses are also shown to vary along the thickness. The results show that as the size of the hole gets smaller and approaches the thickness dimension, the stresses show a marked difference from previously known elastic solutions. In practice, a structural geometry may involve a countersunk rivet hole. The maximum stresses at such holes were recently reported by Shivakumar et al.[3] based on a three-dimensional finite element method. The stress concentration factors for the variation in the countersink angle (Figure 5) and depth of the counter sunk (Figure 6) are presented. The effects of such stresses on the initiation of the flaws as well as propagation of cracks are found to be significant.

III. FLIGHT SPECTRA - STRESS HISTORY

Flight loads are random in nature. For the purpose of analysis and conducting experiments, these loads are transformed into a simplified format. The load information for aircraft structures is obtained in the form of an exceedance diagram (Fig. 7). To develop flight spectra and stress history, the magnitude and the range of accelerations of the center of gravity at a particular location in an aircraft are recorded. Details for stress spectra development are given in MIL-A-8866B[4]. The correlation function between the measured accelerations and stresses depend upon the aircraft and the aircraft configuration. The stress history is a representation of the magnitude and sequence of stresses occurring at a location in the structure. Fuselage structures are subjected to torsional and bending loads due to maneuvering and gust loads on the control surfaces, aerodynamic loads and pressurization cycles. The exceedance diagram in Figure 7 shows the number of times an acceleration range is exceeded. The exceedance diagram shows that the high occuring loads are less frequent. The clipping of these high loads for fatigue analysis is usually done at a level of approximately 10 exceedances per 1000 flights as shown in Fig. 7. Similarly, the lowest loads which occur quite frequently and determine the number of cycles in a spectrum are truncated. The truncation level is normally selected at .1 to .5 million exceedances per 1000 flights. The shaded areas in the diagram must be equal for an acceptable load spectrum.

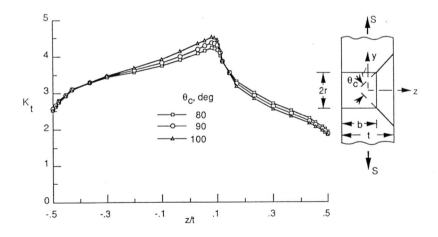

Fig. 5. Variation of Stress Concentration Factor in a C'Sunk Rivet Hole

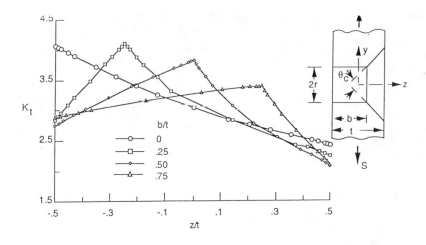

Fig. 6. Stress Concentration Factor Variation with the Depth of the
C'Sunk in Rivet Hole

Deterministic loads are placed in the order, they occur during service. However, probabilistic loads such as those due to gust or maneuvers are arbitrarily sequenced. For a fighter, a low-high-low sequencing per flight is used. A load spectrum normally represents about 1000 flight hours. For proper sequencing of missions, the most severe missions are

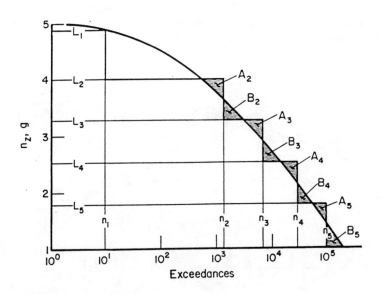

Fig. 7. Construction of Flight Load Spectrum from Exceedances

evenly spaced among other missions. The analysis is usually conducted using the number of load blocks. The block size is generally selected which results in approximately five percent of the growth of an existing crack size.

Due to the nature of flight loads, it is necessary to establish the definition of a load cycle. There are approximately eight different cycle counting methods proposed and used in fatigue analysis of aerospace structures. The method of range-pair counting and the method of rain-flow have been found to be most accurate. The rain-flow method is based on the closed hysteresis loop formed during cyclic loads. A block loading to illustrate the rain-flow cycle counting method is shown in Fig. 8. The constructed rain-flow load cycles for the load block in Figure 8 are listed in Table 1. The load levels are indicated. The rain flows from a load peak to the adjacent peak until the rain starts to drop. It may be noted that the rain can flow only once on a given surface. In the range-pair counting method, a load cycle is counted if, for an ascending load step, the load at level 3 is greater than that at level 1 and if, the load at level 4, is greater than the load at level 2 as shown in Fig. 9. With 2-3 as a load cycle, point 4 becomes 2, point 5 becomes 3, and so on. For a descending load step, the reverse relation between load peak levels apply. After definition of a load cycle i.e., 2 - 3, the point 4 becomes 2 and the next two points are joined to make the required four points. If a load cycle can not be defined, the 1-2 load step become a part of the residue trace. Further details of the range-pair cycle counting method are discussed in Reference 4. The analytic studies have shown that if cycles are not defined or counted, the life prediction is generally unsafe. For loads encountered by aerospace

Table 1. Rain-flow Cycle Analysis of Block Loading (Figure 8)

Count No.	Rainflow Track	Load Range	No. of Cycles
1	1-14	260 KN	1
2	3-4	100 KN	1
3	5-6	100 KN	1
4	7-8	100 KN	1
5	9-14	150 KN	1
6	11-12	100 KN	1
7	13-14	100 KN	1
8	15-18	100 KN	1
9	17-18	90 KN	1
10	19-24	100 KN	1
11	21-24	120 KN	1
12	23-24	90 KN	1

structures, the difference in the prediction of life using the range-pair or the rain-flow method have been found to be small[4]. An important consideration in the construction of load spectrum, load blocks or cycle definition is a firm understanding of the effect of variable loads on damage.

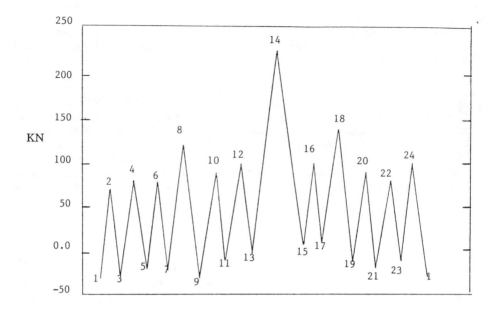

Fig. 8. Block Loading to Illustrate the Rain-flow Cycle Counting Method

IV. SPECTRUM INTERACTION EFFECTS

Life prediction of aerospace structures becomes complicated due to spectrum loads and crack growth, environment and crack growth interaction effects. The experimental data on crack growth under variable loads has established that tensile overloads can cause retardation and compressive underloads can cause acceleration of crack growth. These effects are shown to be enhanced or reduced by the magnitude, frequency and the order of occurrence of variable loads. The crack growth interaction effects are also observed during constant amplitude loading if the load range changes.

Some of the proposed mechanisms responsible for the effects of variable loads on crack growth have been discussed in Reference 14. The crack tip reversed plastic zone, crack closure, crack tip sharpening or blunting, crack tip prestrain, crack tip strain-hardening, mean stress and crack tip dislocation structure have been proposed to explain the crack growth behavior under load variations. The plastic deformation zone due to an overload and accompanying compressive residual stresses in the crack tip region has been used to compensate for the retardation effects in several models[4]. Although such methods are semi-empirical and zone sizes are only estimates, the approaches use adjustment of the value of the fracture parameter which is derived from the crack tip stress analysis. The retardation model most widely used by aerospace designers was originally proposed by Willenborg and has been modified[4,5] for applications to other structural materials. The retardation effect is incorporated by reducing the ratio of the applied stress intensity factors. The applied stress intensity factors are reduced as follows:

$$K_{effective} = K_{applied} - K_{red} \qquad (2)$$

where,

$$K_{red} = K_{max}^{OL} \left[1 - \frac{\Delta a}{Z^{OL}} \right]^{\frac{1}{2}} - K_{max} \qquad (3)$$

The reduction in "K" depends on the maximum value of the stress intensity factor due to an overload, instantaneous crack size, and the plastic deformation zone formed due to the overload. Theoretically, the original model predicts complete retardation at an overload ratio of 2. However, experimental observations of crack growth in various materials show that the overload ratio to shut-off the growth of the crack can be much higher[6]. In the generalized version of the model, this problem has been overcome by introducing an overload shut-off ratio measured experimentally and the threshold value of the stress intensity factor below which the crack grows at a negligible rate. The generalized form of the model is given by:

$$generalized \ K_{red} = \phi \, K_{red} \qquad (4)$$

where,

$$\phi = \left[1 - \frac{K_{max}^{TH}}{K_{max}} \right] [R_{so} - 1]^{-1} \qquad (5)$$

The overload plastic zone size is estimated from the maximum value of the stress intensity factor and the yield strength by the following relation:

$$Z^{OL} = \frac{1}{2\pi} \left[\frac{K_{max}}{\sigma_y} \right]^2 \qquad (6)$$

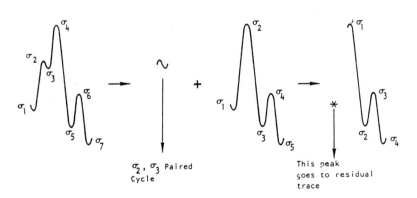

Fig. 9. Range-Pair Method of Cycle Counting

The relation for the size of the plastic zone given by Equation 6 comes from the elastic solution of the crack tip stress field. The stress intensity factor solutions are also calculated assuming elastic conditions. It has therefore been assumed that the plastic deformations under load cycling remains small to permit these assumptions. A critical combination of the stress-strain field is responsible for fracture of the material volume ahead of the crack tip. The plastic deformation under crack tip plane stress is taken to be three times that under plane strain i.e.,

$$Z^{OL}\big|_{plane\ strain} = \frac{1}{3} Z^{OL}\big|_{plane\ stress} \tag{7}$$

$$(\Delta K)_{effective} = K^{max}_{effective} - K^{min}_{effective} \tag{8}$$

$$R_{effective} = \frac{K^{min}_{eff}}{K^{max}_{eff}} \tag{9}$$

The generalized version of Willenborg model can account for the retardation effects for aluminum structures at room temperature quite well. However, the enhanced retardation due to multiple overloads, reduced retardation due to a following compressive load, acceleration of crack growth due to compressive loads and low-high-low tension load combinations are some of the effects still not modeled. An empirical method for the crack acceleration under compression is proposed by Chang[4,5] but still remains to be verified for other materials. In recent years, the methods based on closure of the crack due to unloading has been proposed as discussed by various authors[7-15], perhaps due to its ability to model the physical crack behavior.

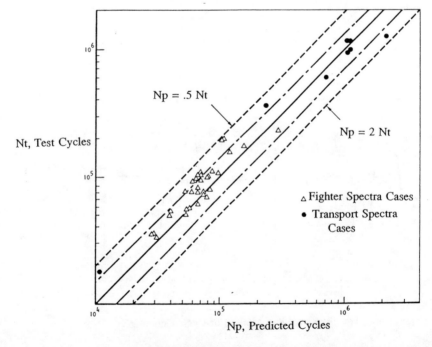

Fig. 10. Prediction of Load Interaction Effects with Residual
Plasticity Model

In a recent survey by Kumar[9], it is shown that the accuracy of life prediction using closure analysis depend on the material properties, specimen geometry, thickness, prestrain, environmental effects and the crack measurement techniques. The effect of thickness on closure was modeled by Clayton[15] based on empirical plastic constraint and stress relaxation.

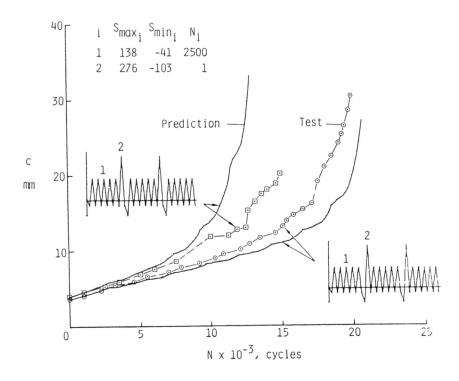

Fog. 11. Prediction of Load Interaction Effects with Closure Model

Newman[10] has proposed a model for tensile overloads based on plasticity induced closure. However, this model does not predict the acceleration effects such as those due to compressive underloads and other associated shortcomings of the closure concept remain unresolved. Plasticity induced closure of cracks is perhaps most dominant under flight induced loads on structures and such a closure mechanism has been able to explain the short crack and overload effects on crack growth accetably. The closure based concepts are still under development for structural applications.

It has been observed that in some load spectra, depending on maneuvers in missions, both effects may be present (i.e., retardation of crack growth and acceleration of crack growth) and the net effect on life is small i.e., the two opposing effects cancel each other. Thus the effects of variable loads depend upon the particular application. It may be noted that some of the assumptions made to develop various crack growth and life models, often do not apply to aerospace structures during service. For example the structural holes may be filled with interference fit fasteners, may be cold worked and may be surface treated to varying degrees.

These treatments have shown to enhance structural life by an order of magnitude. Thus the available models are usually not applicable without some modifications for real life predictions[16]. The experimental data in Figure 10 show that the life predictions using the generalized Willenborg model varies from about half to twice the crack growth life measured during verification tests. The test data was obtained by Chang et al.[4]. For comparison, the predictions of structural life using the closure model are shown in Figure 11. The data show the differences in predicted life and the test life under an overload - underload interruption of the constant amplitude cyclic loading. Further results are described in Reference 12. The reliability of crack growth predictions is much higher over predictions of life for crack initiation. However, further improvement in life prediction methods are required to assure safety and improved operations.

V. EQUIVALENT DAMAGE CONCEPT

The equivalent damage concept is based on the idea that the life of a large structure can be predicted from a small representative volume of structural material by equating damage in a large structure to damage in a small volume. An equivalent value of the fracture or the damage parameter is determined via test data from structural components, structural elements and material coupons. This concept is schematically described in Figure 12. An equivalent variable load block is developed from structural component tests conducted under flight spectrum loads.

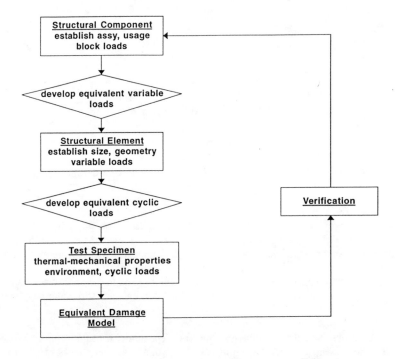

Figure 12. An Equivalent Damage Model Development Scheme

96

Further tests on structural elements are conducted to establish the effects of size, geometry and variable loads. From these tests, the equivalent cyclic loads may be developed. These loads may be reduced to constant amplitude cyclic loads by incorporating the effects of the mean load or the stress ratio on crack growth. In simplistic form, the general relation may take the form:

$$\left[\tilde{D} \right] = c \left[\tilde{D}_p \right]^n \tag{10}$$

The above equation relates the occuring damage rate D with the damage parameter D_p. The symbols c and n are the test correlation constants. Such forms of relations have previously been

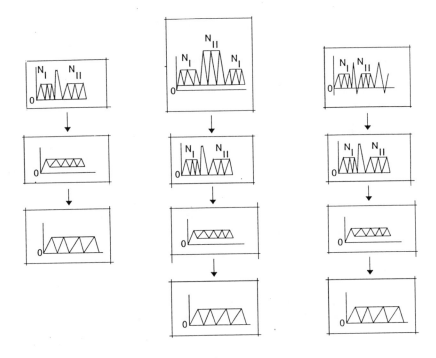

Figure 13. Transformation of Variable Load Cycles to Constant Amplitude Load Cycles

used to describe the effects of plasticity, creep, fatigue and fracture for both the time dependent and time independent phenomena. From a mathematical modeling point of view, such a function is very useful since the scenarios of zero damage, constant damage, linear or non-linear damage can be modeled using such a relation simply by changing the values of the variables. For these reasons, this form is selected for equivalent damage modeling. With such a scheme an arbitrarily varying load with mechanical and thermal components which vary with time can be reduced to a simple load block such as that shown in Figure 2. Using the equivalent damage concept, the load blocks can be further reduced to cyclic loads with a fixed mean and amplitude (i.e., an equivalent stress ratio). The expression then for the damage due to a growing crack may be written as:

$$\left[\ \widetilde{D} \ \right] = c \left[\ \widetilde{K} \ \right]^{n} \tag{11}$$

where,

$$\widetilde{K} = K_{max} f_{R}(R) \tag{12}$$

In the above equations, the damage parameter is replaced by the stress intensity factor. The cyclic loads are further reduced to loads involving only the maximum stress intensity factor and in the final form, they are independent of the stress ratio:

$$\frac{da}{dN} = c \ K_{max}^{n} \tag{13}$$

The value of f(R) as a function of R can be obtained from cyclic crack growth tests. This stress ratio function can be used for crack growth analysis involving stress ratios. The equivalency of damage at other levels of the structure can similarly be made by using test data under block loads, variable loads and cyclic loads.

A schematic illustration of transforming variable loads to equivalent constant amplitude cyclic loads is presented in Figure 13. Various forms of load variations such as overloads, underloads, sustained loads, multiple loads, tensile-compressive loads can be reduced to a single load block with an overload or an underload and a number of constant amplitude cycles. Ultimately, all loads are transformed to cyclic loads involving only the maximum stress intensity factor. When the crack length is constant, the maximum stress intensity factor becomes the maximum stress in Equation 13 and may be representative of crack re-initiation. Such a relation for crack re-initiation is the same as that for fatigue crack growth.

VI. SUMMARY AND CONCLUSIONS

The life analysis methodology for aerospace structural design is reviewed. Recent developments in the area of stress and fracture mechanics analysis of plates with notches including circular and countersunk rivet holes are presented. Fracture mechanics based models for fatigue crack growth and spectrum load interaction effects are discussed. An equivalent damage concept for life analysis is discussed. Based on equivalent damage, the proposed scheme provides a method of transforming the variable loads in a spectrum to simplified constant amplitude load cycles. A use of interference fit fasteners combined with cold worked holes has shown to increase the structural life significantly. Advanced structures with thermal loads may involve more complicated interaction and coupling effects. The life prediction methods for such applications are currently under development.

ACKNOWLEDGEMENT

The author expresses his sincere thanks to Mr. Joseph Burns, Fatigue, Fracture and Reliability Section, Structural Integrity Branch, Wright Laboratory for his review of this paper and making several useful comments.

REFERENCES

1. J.C. Newman and I.S. Raju, "Stress Intensity Factor Equations for Cracks in Three Dimensional Finite Bodies Subjected to Tension and Bending Loads", NASA Technical Memorandum, 85793, 1984.

2. E.S. Folias and J.J. Wang, "On the Three-Dimensional Stress Field Around a Circular Hole in a Plate of Arbitrary Thickness", Computational Mechanics, Vol. 6, No. 3, 1990.

3. K.N. Shivakumar and J.C. Newman, Jr., "Stress Concentrations for Straight-Shank and Countersunk Holes in Plates Subjected to Tension, Bending and Pin Loading", NASA Technical Paper 3192, June 1992.

4. J.B. Chang, R.M Hiyama and M. Szamossi, "Improved Methods for Predicting Spectrum Loading Effects," AFWAL-TR-81-3092, Volume 1, November 1981.

5. D.M. Harmon and C.R. Saff, "Damage Tolerance Analysis for Manned Hypervelocity Vehicles," WRDC-TR-89-3067, 1989.

6. A. Nagar, "Fatigue Crack Growth Retardation in Ti 6242S", 35th Structures, Structural Dynamics and Materials Conference, Hilton Head, SC, 1994, to appear.

7. K. Ohji, K. Ogura and Y. Ohkubo, "Cyclic Analysis of Propagating Crack and its Correlation with Fatigue Crack Growth", Engineering Fracture Mechanics, Volume 7, 1975.

8. J.C. Newman and I.S. Raju, "Prediction of Fatigue Crack Growth Patterns and Life in Three Dimensional Cracked Bodies", Sixth International Conference on Fracture, New Delhi, India, 1984.

9. R. Kumar, "A Review on Crack Closure for Single Overload, Programmed and Blocked Loadings", Engineering Fracture Mechanics, Volume 42(1-3), 1992.

10. J.C. Newman, "Prediction of Fatigue Crack Growth Under Variable Amplitude and Spectrum Loading Using a Closure Model", ASTM STP 761, American Society for Testing and Materials, 1982.

11. S.G. Russell, "A New Model for Fatigue Crack Growth Retardation Following an Overload", Engineering Fracture Mechanics Volume 33, No. 6, 1989.

12. R.P. Wei, "Fatigue Crack Growth Response Following a High Load Excursion in Aluminum Alloy", Journal of Engineering Materials and Technology, Transactions of the ASME, Vol. 102, 1980.

13. J. Schijve, "Fatigue Crack Growth Predictions for Variable Amplitude Loading and Spectrum Loading", Delft University of Technology, Report LR-526, 1987.

14. K.N. Raju and R. Sunder, "Unified Approach for Modeling Fatigue Crack Growth and Some Observations of Behavior Under Flight Simulation Loading", International Conference on Fracture, New Delhi, India, 1984.

15. J.Q. Clayton, "Modeling Delay and Thickness Effects in Fatigue", Engineering Fracture Mechanics, Volume 32, No.2, 1989.

16. M.A. Landy, H. Armen and H.L. Eidinoff, "Endurance Stop-Drill Repair Procedures for Cracked Structure", ASTM Symposium on Fatigue in Mechanically Fastened Joints, Charleston, S. Carolina, 1985.

APPENDIX - 1

Stress Intensity Factor Solution for a Single Corner Crack from a Central Open Hole

(Reference 5)

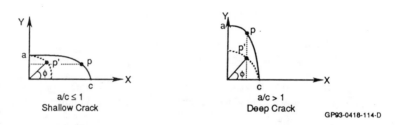

GP93-0418-114-D

The angle, ϕ, represents the point along the part-through crack perimeter, P, at which the stress intensity, K, is computed.

To calculate K for the crack depth, a, $\phi = 90$
To calculate K for the crack length, c, $\phi = 10$

The point, P, is defined using the following procedure:

1) A circular arc is drawn which is within the crack perimeter and which has a radius equal to "a" for a shallow crack (or "c" for a deep crack);

2) The point along the arc, P', is defined by ϕ which is measured from the X axis;

3) A line is drawn parallel to the X axis for a shallow crack (or parallel to the Y axis for a deep crack) through P' to the crack perimeter.

$$K = K_t(\phi) + K_b(\phi)$$

$$t - \text{tension}, \ b - \text{bending}$$

$$K_a = K_t(\phi=90^\circ) + K_b(\phi=90^\circ)$$

$$K_c = K_t(\phi=10^\circ) + K_b(\phi=10^\circ)$$

$$K_t(\phi) = \sigma_t \sqrt{\pi \frac{a}{Q}} \, F \qquad\qquad K_b(\phi) = \sigma_b \sqrt{\pi \frac{a}{Q}} \, F \, H$$

where

$$Q = 1 + 1.464 \left(\frac{a}{c}\right)^{1.65} \qquad \text{for} \quad \frac{a}{c} \leq 1$$

$$Q = 1 + 1.464 \left(\frac{a}{c}\right)^{1.65} \qquad \text{for} \quad \frac{a}{c} > 1$$

$$F = m \, g_1 \, g_2 \, g_3 \, g_4 \, f_\phi \, f_w$$

$$H = H_1 + (H_2 - H_1) \sin^p \phi$$

and

$$g_2 = \frac{1 + 0.358\lambda + 1.425\lambda^2 - 1.578\lambda^3 + 2.156\lambda^4}{1 + 0.13\lambda^2}$$

where

$$\lambda = \frac{1}{1 + c/R \, \cos(\mu\phi)}$$

$$\mu = 0.85 \qquad\qquad\qquad\qquad\quad - \text{tension}$$

$$\mu = 0.85 - 0.25 \, (a/t)^{1/4} \quad - \text{bending}$$

$$f_w = \left[\sec\left(\frac{\pi R}{2b}\right) \sec\left(\frac{\pi(2R+nc)}{4(b-c)+2nc} \sqrt{\frac{a}{t}}\right) \right]^{1/2} \qquad \begin{array}{l} n = \text{\# of cracks} \\ (1 \text{ or } 2) \end{array}$$

101

AD-Vol. 36, Fatigue and Fracture of Aerospace Structural Materials
ASME 1993

THERMAL STRESS INTENSITY FACTOR OF A
CIRCUMFERENTIALLY CRACKED BI-METALLIC TUBE

An-Yu Kuo
Structural Integrity Associates, Inc.
San Jose, California

ABSTRACT

A modified J-integral formulation was derived and used to calculate stress intensity factors of a bi-metallic tube with a circumferential crack. To extract the effects of the residual stresses resulting from material properties mismatch on the thermal stress intensity factor, a cracked clad tube under a constant temperature was studied in this paper. It was found that the stress intensity factor would reach a maximum value at a very shallow crack and rapidly approach an almost constant value as the crack grows deeper. The modified J-integral formulation was proven to be an accurate and convenient way of computing thermal stress intensity factors.

1.0 INTRODUCTION

It is a common practice to clad the inside of a cylinder or tube with a different material of more corrosion resistance. For example, stainless steels have been used as clad to the inside of reactor pressure vessels made of low alloy steel in the power and petro-chemical industries. Due to the mismatch in coefficients of thermal expansion (CTE) between the clad and base materials, a very high "residual" stress may exist even when the tube or cylinder is not under any mechanical loading. The term "residual" stress used in this paper is actually the difference of thermal stresses between the bi-metallic tube and a single material tube of the same size and under the same temperature distribution. Such a high "residual" stress often result in initiation and propagation of crack at the inside surface of the clad. The "residual" stress also affects the stress intensity factor of the crack. This paper studies the stress intensity factor of a circumferential crack in a bi-metallic tube due to the "residual stress".

2.0 PROBLEM DESCRIPTION AND ASSUMPTIONS

For simplicity, it is assumed that (a) both the clad and base materials are linear elastic and isotropic, and (b) the tube is long enough that it can be treated as an infinitely long cylinder.

As illustrated in Figure 1, a bi-metallic tube with outside radius R_o, inside radius R_i, clad radius R_c, and a circumferential crack depth of a is studied. Under thermal loading, the thermal stress in the bi-metallic tube will arise from two sources: thermal stress due to the through-wall temperature gradient and the "residual" stress. In other words, as illustrated in Figure 2, when the bi-metallic tube is under loading, there may exist an average wall temperature change ΔT_m from the stress-free temperature and a through-wall temperature gradient of approximately $\dfrac{\Delta T_o - \Delta T_i}{R_o - R_i}$, where ΔT_o and ΔT_i are temperature changes at the outside and inside surfaces, respectively. It is well understood that the through-wall temperature gradient may cause thermal stress and, thus, stress intensity factor if the tube is cracked. For an unrestrained single material tube, the thermal stress as well as the stress intensity factor due to the average temperature change ΔT_m will both be zero. Conversely, when the cracked bi-metallic tube, shown in Figure 1, is under a uniform temperature change ΔT_m, the thermal stress and stress intensity factor will not be zero because a self-equilibrium but very high "residual" stress, as depicted in Figure 2, would develop. In this paper, it is assumed that the circumferentially cracked bi-metallic tube, as illustrated in Figure 1, is subject to a uniform temperature change of ΔT_m. However, the analytical and numerical methods described in this paper are also applicable to any through-wall temperature distribution.

3.0   METHODOLOGY

The J-integral has been widely used by fracture mechanics researchers in various applications. One application of the J-integral is to calculate stress intensity factors based on the following relationship:

$$J = \frac{1-v^2}{E}(K_I^2 + K_{II}^2) + \frac{1+v}{E}K_{III}^2 \tag{1}$$

where K_I, K_{II}, and K_{III} are stress intensity factors of the three fracture modes. For a cracked linear elastic solid under the plane strain condition and a steady-state temperature distribution, it has been shown by many researchers (e.g., Wilson and Yu, 1979; and Aoki, et al., 1981) that an area integration might be necessary in the J-integral formulation:

$$J = \int_\Gamma (Wn_1 - T_i u_{j,1})ds + \int_A \sigma_{ij}\delta_{ij}(1+v)\alpha\theta_{,1}dA \tag{2}$$

where Γ is an integration contour enclosing the crack-tip, $W = \frac{1}{2}\sigma_{ij}[e_{ij} - \delta_{ij}\alpha(1+v)\theta]$ is the strain energy density, θ is the temperature distribution within the solid, T_i is the traction, n_i is the i-th component of the outer normal along Γ, and A is the area enclosed by the integration contour. The area integration in Eq. (2) is undesirable because it not only make the numerical calculation more complex, but also requires a finer finite element mesh near the crack-tip. It has been shown by Kuo and Riccardella (1987) that the area integration of the J-integral formulation in Eq. (2) can be avoided by using the following formulation:

$$J = \int_\Gamma [Wn_1 - T_i u_{i,1} + \alpha(1+v)(T_1\theta + T_2\Omega)]ds \tag{3}$$

where Ω is a complementary temperature distribution determined by:

$$\theta_{,1} = \Omega_{,2} \quad , \quad \theta_{,2} = -\Omega_{,1} \tag{4}$$

J-integral formulations for axisymmetric solids are different from the above formulations for the plane strain condition. For an axisymmetric solid under mechanical loading, the following J-integral formulation has been derived by Bergkvist and Lan Huong (1977) and Astiz, et al., (1977):

$$J = \frac{1}{R_a}\{\int_\Gamma [Wrdz-(T_r u_{r,r}+T_z u_{z,r})rds]-\int_A (W-T_\phi u_r/r)dA\} \tag{5}$$

where r, z, and ϕ are the cylindrical coordinates, and R_a is the radius of the crack-tip as depicted in Figure 1. The area integration in Eq. (5) was eliminated by Kuo (1987) by the following formulation:

$$J = \frac{1}{R_a^2} \int_\Gamma [Wx_p n_p - T_p u_{p,q} x_q -\frac{1}{2}T_p u_p]ds \tag{6}$$

where the subscripts p and q are either the r- or z-coordinate of the polar coordinate system.

The J-integral formulation for an axisymmetric solid under a steady-state temperature distribution can be derived with the same method described in Kuo and Riccardella (1987) and Kuo (1987). After some mathematical manipulation, the following J-integral formulation can be derived for the axisymmetric solid under thermal loading:

$$J = \frac{1}{2\pi R_a^2} \{\int_\Gamma [Wx_p n_p-T_p u_{p,q} x_q -\frac{1}{2}T_p u_p]rds$$
$$+ \int_A \alpha(\sigma_{rr}+\sigma_{zz}+\sigma_{\phi\phi})(\theta_{,p} x_p+\frac{3}{2}\theta)rdA\} \tag{7}$$

where p,q=r or z, x_r=r, x_z=z, and A is the area enclosed by the integration contour Γ. In the above derivation (Kuo, 1987), the M-integral and the relationship between M and J

$$M = 2 \pi R_a^2 J \tag{8}$$

has been used. It is seen in Eq. (7) that the area integration term can no longer be avoided from the J-integral formulation. The additional area integration term in the J-integral makes it necessary to use fine finite element mesh at the crack-tip.

105

4.0 EXAMPLE PROBLEM AND DISCUSSION

A reactor pressure vessel of the following dimensions and material properties is chosen as the example component:

R_o	62.375	inches
R_c	54.5	inches
R_i	54.391	inches
E	29,000	ksi (base material)
E	28,000	ksi (clad material)
α	7.47E-6	in/in-°F (base material)
α	9.50E-6	in/in-°F (clad material)
ν	0.3	

Uniform Tension in a Cracked Cylinder

To check the accuracy of the J-integral formulation in Eq. (7), a uniform remote tension of 1 ksi was first applied and a circumferential crack depth of a=2.6465 inches was assumed. A two-dimensional finite element program was used to calculate the J-integral. To assure the finite element mesh being fine enough at the crack-tip, a total of 25 elements were used in the thickness while the aspect ratios of the elements were kept very close to 1 in the crack-tip region. The resulting stress intensity factor was also compared with solutions from a handbook (Tada and Paris, 1985) and from ABAQUS (1989). The K solutions are tabulated below:

Present	4.04 $ksi\sqrt{in}$
Handbook	4.05 $ksi\sqrt{in}$
ABAQUS	4.04 $ksi\sqrt{in}$

The three solutions are in perfect agreement.

A Penny-Shaped Crack in an Infinite Solid with a Constant Temperature at the Crack Surfaces

The second problem considered is a penny-shaped crack with a radius of 8 inches in an infinite solid of E=29,000 ksi and α =7.47E-6 in/in-°F. The crack surface temperature was kept at 1°F while temperatures at infinity were at 0°F. Again, the J-integral solution was calculated by the two-dimensional finite element program. A solid bar of 80 inches in radius and 160 inches long was used to simulate the infinite solid. Resulting K solutions from the present study and the other two reference methods are as follows:

Present	0.47 $ksi\sqrt{in}$
Handbook	0.49 $ksi\sqrt{in}$
ABAQUS	0.47 $ksi\sqrt{in}$

Again, the present solution is in perfect agreement with the ABAQUS solution and is within 5% of the handbook solution. The 5% difference is probably due to the fact that a finite cylinder was used to simulate the infinite domain.

A Circumferentially Cracked Bi-Metallic Cylinder Under a Uniform Temperature Rise

The same clad cylinder described in the first example was studied again in the third example. The entire bi-metallic cylinder was assumed to have a temperature drop of 289°F, i.e., ΔT_m =-289°F. K solutions for crack depths up to 80% of the wall thickness were calculated and depicted in Figure 4. It is seen from this figure that the stress intensity will not be zero even when the entire cylinder is unconstrained under a uniform temperature. It is well known that the thermal stress and stress intensity factor would be zero if the cylinder is in a single material. The stress intensity factors shown in Figure 4 are caused by the "residual" stress in the bi-metallic cylinder. It is also seen from Figure 4 that the stress intensity factors reach a very high maximum at a crack depth of the clad thickness and quickly decay to an almost constant value as the crack grows deeper. The quick drop in the stress intensity solutions is typical for displacement-controlled fracture. As illustrated in Figure 5, bending moments across the remaining ligament of the cracked pipe were also calculated. The bending moment at the remaining ligament is the driving force of the crack. It is seen from Figure 5 that the bending moments at the remaining ligament decrease almost linearly with crack depths. The quick drop in stress intensity factors in Figure 4 can explain why many clad cylinders and tubes would exhibit many shallow cracks in the inner clad, but seldom have cracks extending deep into the base material, unless there are other severe thermal or mechanical loads. It is worth noting that, although a constant through-wall temperature was assumed for this example problem, the method also applies to any arbitrary temperature distributions.

5.0   CONCLUSIONS

The following conclusions can be drawn from this study:

1. The J-integral formulation of Eq. (7) provides an accurate and simple way of calculating stress intensity factors for circumferentially cracked tubes.

2. The "residual" stress and stress intensity factor will not be zero for a bi-metallic tube or cylinder even when the tube or cylinder is at a constant temperature.

3. The stress intensity factor due to the "residual" stress would reach a very high maximum value, when the crack is at the material interface, but would quickly drop to a much lower, constant value as the crack grows deeper.

6.0   ACKNOWLEDGEMENT

The support and review provided by Mr. Bob Carter of Yankee Atomic Electric Company is appreciated.

7.0   REFERENCES

ABAQUS User's Manual, Version 4.0, 1989, Hibitt-Karlsson-Sorensen Inc.

Aoki, S., Kishimoto, K., and Sakata, M., 1982, "Elastic-Plastic Analysis of Crack in Thermally-Loaded Structures," Engineering Fracture Mechanics, Vol. 16, pp. 405-413.

Bergkvist, H., and Lan Huong, G. L., 1977, "J-Integral Related Quantities in Axisymmetric Cases", International Journal of Fracture, Vol. 13, pp. 556-558.

Astiz, M.A., Elices, M., and Galrez, V.S., 1977, "On Energy Release Rates", Fracture 1977, Vol. 3, ICF4, Waterloo, Canada, June 19-24, pp. 395-399.

Kuo, A. Y., 1987, "On the Use of a Path-Independent Line Integral for Axisymmetric Cracks with Non-Axisymmetric Loading," Journal of Applied Mechanics, Vol. 54, pp. 833-837.

Kuo, A. Y., and Riccardella, P. C., 1987, "Path-Independent Line Integrals for Steady-State, Two-Dimensional Thermoelasticity," International Journal of Fracture, Vol. 35, pp. 71-79.

Tada, H, Paris, P. C., and Irwin, G. R., 1985, "The Stress Analysis of Cracks Handbook," second edition, Paris Products Inc.

Wilson, W. K., and Yu, I. W., 1979, "The Use of J-Integral in Thermal Stress Crack Problems", International Journal of Fracture, Vol. 15, pp. 377-387.

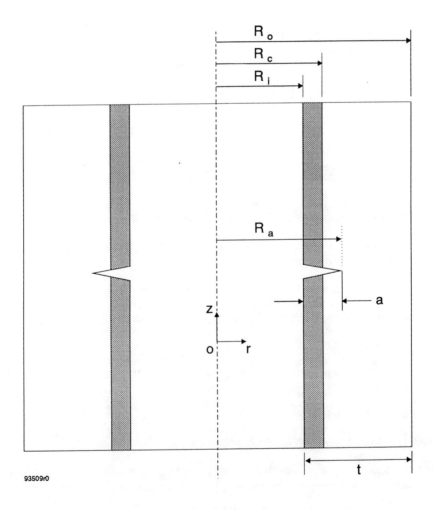

Figure 1. A Bi-Metallic Tube with a Circumferential Crack

93509r0

108

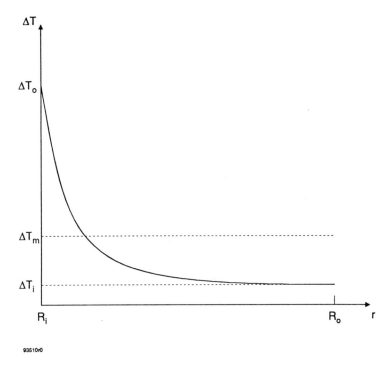

Figure 2. A Through-Wall Temperature Change

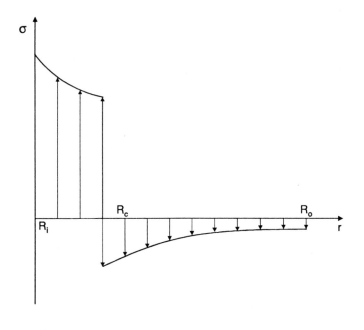

Figure 3. A Typical "Residual" Stress Distribution in a Bi-Metallic Tube

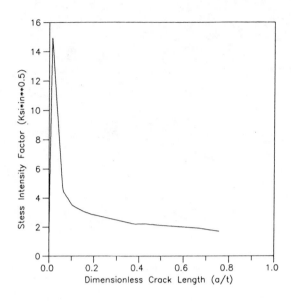

Figure 4. Resulting Stress Intensity Factors of a Cracked Bi-Metallic Cylinder

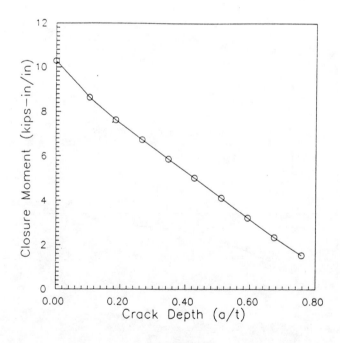

Figure 5. Moment Reduction Versus Crack Depths

110

AD-Vol. 36, Fatigue and Fracture of Aerospace Structural Materials
ASME 1993

ANALYSIS OF TWO-DIMENSIONAL INTERACTING CRACK PROBLEMS USING A FINITE ELEMENT ALTERNATING METHOD

T. S. Moore and R. Greif
Department of Mechanical Engineering
Tufts University
Medford, Massachusetts

C. Mauge
Radionics Incorporated
Burlington, Massachusetts

1 Abstract

A finite element alternating method (FEAM) is developed for the solution of bounded two dimensional elastic problems with multiple cracks. A computer code is developed to predict the effects of multiple crack interaction as well as crack/boundary effects.The code combines the analytical crack interaction technique developed by Kachanov for cracks in infinite domains, with finite element methods (in conjunction with the Schwarz-Neumann alternating technique),to establish the solution in a finite domain. The combination of these two solutions in an iterative procedure produces the finite element alternating method for a finite plate. The main objective of the current work is to investigate the feasibility, efficiency and accuracy of the FEAM using the Kachanov technique for the analytic part of the solution.

2 Introduction

Problems relating to interacting cracks and their effect on crack linkup and propagation are important in predicting damage tolerance. One field particularly dependent on investigations of this type involves the structural integrity of aging airplanes[1]. As aircraft are increasingly used past their design life, fatigue cracks develop which must be analyzed carefully and carefully monitored with sophisticated inspection procedures. Structures of this type often have complicating configurations such as finite boundaries, stiffeners, rivets and cutouts. The finite element alternating method is an efficient method for analyzing complicated configurations containing multiple cracks.

This investigation involves combining the Kachanov method[2] for multiple cracks in *infinite* domains with the finite element method, to establish the solution for the problem of multiple cracks in a *finite domain*. The Kachanov method is a simple technique of stress analysis that produces analytical solutions for arbitrarily located and oriented cracks, which are accurate up to quite close spacings between cracks. The key simplifying assumption of the method is that the traction non-uniformities (with zero average) on the i-th crack have negligible impact on the k-th crack. This assumption may be considered as an analogue of Saint Venant's principle for crack interaction problems. The finite element method (FEM) is used in conjunction with the Schwarz-Neumann alternating technique to solve the problem. In the current work, 4 noded isoparametric elements with 3 x 3 lattice of integration are used.

In the alternating method the boundary conditions are satisfied by alternating (iterating) between two solutions. The first solution uses the Kachanov method for the cracked infinite plate subjected to surface stresses, to predict the stress everywhere. The second solution uses the FEM in which the forcing functions are the opposite of the stresses generated by the first solution at the appropriate points along the finite boundary. The combination of these two solutions in an iterative procedure produces the finite element alternating method FEAM for the finite plate. The main objective of the current work is to investigate the feasibility, efficiency and accuracy of the FEAM using the Kachanov technique for the analytic part of the solution.[3]

3 The Kachanov Technique

The *Kachanov Method* is a simple technique of stress analysis which produces accurate results for elastic bodies where multiple cracks and their interactions are to be considered. The technique of superposition and the idea of self-consistency applied to the average tractions on individual cracks form the basis for this method[2].

If we replace the case of a linear elastic body with N cracks (unit normals $\mathbf{n_i}$) under a remote loading $\boldsymbol{\sigma}^\infty$ by the equivalent problem of the cracks with faces loaded by tractions $-\mathbf{n_i}\cdot\boldsymbol{\sigma}^\infty$ and stresses vanishing at infinity, then we can represent it as a superposition of N sub-problems, each with one crack loaded with unknown tractions. Examining the ith sub-problem, the traction on crack i is the summation of the traction $-\mathbf{n_i}\cdot\boldsymbol{\sigma}^\infty$ and the tractions $\Delta\mathbf{t_{ki}}(\xi)$ (on any given crack ξ runs $\pm a$) induced on the ith crack line due to the interaction of the other kth cracks. *The simplicity of the Kachanov method lies in the fundamental assumption that* $\Delta\mathbf{t_{ki}}(\xi)$ *is generated by the* kth *crack loaded by its* **uniform average** *traction.* This assumption neglects the impact on the ith crack of the traction non-uniformities $\mathbf{t_k} - \langle\mathbf{t_k}\rangle$ (the notation $\langle\rangle$ represents the average) from the kth crack. In this way we can represent the normal and shear tractions on a typical crack defined as crack 1 as

$$p_1(\xi) = p_1^\infty + \mathbf{n_1} \cdot [\boldsymbol{\sigma}_2^n(\xi)\langle p_2\rangle + \boldsymbol{\sigma}_2^\tau(\xi)\langle\tau_2\rangle\cdots + \boldsymbol{\sigma}_N^n(\xi)\langle p_N\rangle + \boldsymbol{\sigma}_N^\tau(\xi)\langle\tau_N\rangle]\cdot\mathbf{n_1}$$
$$\tau_1(\xi) = \tau_1^\infty + \mathbf{n_1}\cdot[\boldsymbol{\sigma}_2^n(\xi)\langle p_2\rangle + \boldsymbol{\sigma}_2^\tau(\xi)\langle\tau_2\rangle\cdots + \boldsymbol{\sigma}_N^n(\xi)\langle p_N\rangle + \boldsymbol{\sigma}_N^\tau(\xi)\langle\tau_N\rangle]\cdot(\mathbf{I} - \mathbf{n_1 n_1}) \tag{1}$$

where p_1^∞ and τ_1^∞ are the normal and shear tractions induced on the first crack by the remote loading, and $\boldsymbol{\sigma}_i^n, \boldsymbol{\sigma}_i^\tau$ are the stress fields generated by the ith crack with normal (n) and shear (τ) loading. Averaging Equation 1 over the length of crack 1 results in the following

$$\langle p_1\rangle = p_1^\infty + \Lambda_{21}^{nn}\langle p_2\rangle + \Lambda_{21}^{\tau n}\langle\tau_2\rangle + \cdots\Lambda_{N1}^{nn}\langle p_N\rangle + \Lambda_{N1}^{\tau n}\langle\tau_N\rangle$$
$$\langle\tau_1\rangle = \tau_1^\infty + \Lambda_{21}^{n\tau}\langle p_2\rangle + \Lambda_{21}^{\tau\tau}\langle\tau_2\rangle + \cdots\Lambda_{N1}^{n\tau}\langle p_N\rangle + \Lambda_{N1}^{\tau\tau}\langle\tau_N\rangle \tag{2}$$

where Λ are termed the *transmission factors* which characterize the effect of the average tractions from one crack to another. The indices work as follows, $\Lambda_{21}^{n\tau}$ is the average *shear* (τ) on crack *one* caused by the *normal* (n) uniform load on crack 2 [2, 4].

Considering all N cracks, Equation 2 forms a system of 2N linear algebraic equations that can be solved for the average tractions $\langle\mathbf{t}_k\rangle = \langle p_k\rangle, \langle\tau_k\rangle$ as

$$(2\delta_{ik}\mathbf{I} - \boldsymbol{\Lambda}_{ik}) \cdot \langle\mathbf{t}_k\rangle = \mathbf{t}_k^\infty \tag{3}$$

where δ_{ik} is the Kronecker delta. Substituting the average tractions $\langle\mathbf{t}_i\rangle$ into Equation 1 the actual tractions $t_k(\xi)$ are found and used in a Green's function formulation of the point force

112

solution

$$\left\{ \begin{array}{c} K_I \\ K_{II} \end{array} \right\}_{\pm a} = \frac{1}{\sqrt{\pi a}} \int_{-a}^{+a} \left\{ \begin{array}{c} p(\xi) \\ \tau(\xi) \end{array} \right\} \sqrt{\frac{a \pm \xi}{a \mp \xi}} d\xi \qquad (4)$$

to calculate the SIF at the crack tips[2, 4].

This method is quite accurate and compares favorably against other methods such as polynomial approximation techniques. For example, in the case of two collinear cracks with tips in close proximity (distance between inner tips one full order of magnitude smaller than crack length), the error is only 1.4% for the inner and .2% for the outer tips. This technique has been used extensively to generate approximate analytical solutions for the SIF of the many crack problem, accurate up to small inter-crack distances in *infinite* media. The present work extends the Kachanov method [2] to the finite domain.

4   Finite Element Alternating Method

The stress intensity factor (SIF) is an important parameter in the design of safe structures. When the SIF reaches a critical value, K_c, a unique parameter based on the material properties and atomic structure, it is likely that unstable crack propagation will initiate. In the literature several analytical methods are developed to evaluate SIF for idealized crack problems with infinite domains[5, 6]. However for finite crack problems the complicated boundary conditions make analytical solutions difficult to obtain. As a result research in this area has focused on numerical techniques[7] such as FEAM.

In the recent literature[7, 8, 9, 10, 11] the finite element alternating method is shown to be both an accurate and efficient method for the analysis of 2 and 3 dimensional crack geometries. The method is based on the Schwartz-Neumann alternating method and is implemented numerically.

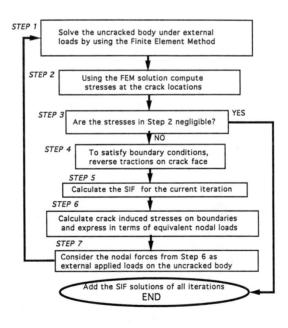

Figure 1: FEAM Procedure

4.1   Basic Procedure

The FEAM alternates between two solutions to satisfy both the crack and finite edge boundary conditions. The solutions necessary for this method are

- Analytical solution for a cracked infinite plate under arbitrary load (Equation 4).

- Numerical stress solution for finite body under arbitrary load. (FEM)

The basic procedure (Figure 1) is as follows [8].

1. Using the FEM, analyze a problem with the same applied loads and geometry as the problem of interest, but *without any cracks.*

2. The FEM yields the stresses throughout the entire body, including the lines coincident with the cracks in the original problem. From the FEM stresses calculate the tractions on these lines.

3. If tractions on the crack faces are negligible (below tolerance) stop procedure and calculate the sum of the SIF computed so far. If tractions are above tolerance go to step 4.

4. To free the tractions on the crack faces, thus satisfying proper crack face boundary conditions, the normal and shear tractions computed in step 2 must be removed. To do this the negative of the crack face tractions are applied in the analytical solution.

5. From the tractions calculated in step 2 the SIF are determined by evaluating them in the Green's function.(Equation 4)

6. The reversed crack face normal and shear tractions from step 4 create tractions on all the boundaries of the problem. These are calculated analytically using the Westergaard technique

$$\sigma_x = ReZ - yImZ'$$
$$\sigma_y = ReZ + yImZ'$$
$$\tau_{xy} = -yReZ'$$

(5)

for mode I, and similar ones for mode II.

7. The boundary tractions are expressed in terms of nodal forces on the edges of the finite element mesh. The new load vector that results is substituted into step 1, and the uncracked body once again is numerically analysed using FEM.

4.2   Computational Aspects of FEAM

This code was implemented in FORTRAN on a Convex main frame. Root FEM routines from the FEABL library (MIT)[12] were used as the foundation for this work. A great deal of effort was spent on the computational aspects of this work. This section deals with some of the computational details involved in the code that are specific to this work.

Element Type – The elements used in the FEM subsection of the method are of the 4-noded isoparametric quad variety. These elements are bilinear in their shape functions and offer a basic level of accuracy. While 8-noded isoparametric quadratic elements were originally intended and would have offered a very marked improvement in accuracy they were unavailable in a form suitable for the programing of this code. However, the level of accuracy available in the 4-noded quad is sufficient to show the feasibility of using the Kachanov Technique with the FEAM.

Stress Generator – The stress generator extracts the elemental displacement vectors, {u}, and applies

$$\{\sigma(r,s)\} = [C][B]\{u\}$$

(6)

at the center point(the 2x2 guassian point) of each element where $[B]$ is the matrix of strain-displacement transformations, and $[C]$ is the matrix of elastic constants. The calculated elemental stresses σ_x, σ_y,and τ_{xy} are placed into data arrays of size NxM, where N is the number of elements in the x-direction and M is the number of elements in the y-direction.

Calculation of Crack Tractions – The stresses on the crack lines are computed at three points; both tips and the center. For the crack sizes, geometries, and loading conditions considered three evaluation points are sufficient for reasonable accuracy. The stress values for these three points are calculated from the stress data arrays using a bivariate quadratic interpolation routine. With these values for stress the equivalent tractions are calculated at the three crack points from standard theory of elasticity,

$$\{t_n\} = [\sigma] \cdot \{n\}$$
$$\sigma_n = \{t_n\}^T \cdot \{n\}$$
$$\{\tau\} = \{t_n\} - \sigma_n\{n\}$$

(7)

where $[\sigma]$ is the stress matrix, $\{n\}$ is the vector normal to the crack, $\{t_n\}$ is the traction on the crack face, σ_n is the stress component normal to the crack, and $\{\tau\}$ is the shear traction on the crack. The value for the shear component is found from the magnitude of $\{\tau\}$ and the sign from the traction direction.

Representation of Tractions as Polynomials and Calculation of Stresses at Boundaries – The subroutine responsible for fitting the three calculated point tractions to a polynomial through a least squares fit, which also containes the FORTRAN call that calculates the stresses at the boundary nodes. The location of each boundary node is identified from a data file. Equation 5 and similar ones for mode II are utilized to calculate the stresses at these points. These values are stored in data files for later use.

Conversion of Stresses into Normal and Shear Components and Calculation of Nodal Forces – The code reads in values for the normal to the boundary at each boundary node and, using a similar procedure to Equation 7, calculates the normal and shear components at each point. The next subroutine then follows FEABL protocol and converts these values to nodal forces to be reapplied to the next finite element analysis. A great deal of attention is paid to the trigonometry involved with these operations. The angles specified for crack normals and normals to the boundaries must be of a consistent nature and the computed values of trigonometric functions in different quadrants needs to be carefully done.

Application of Kachanov Technique The way the FEAM procedure is formulated in the preceding sections, the results for the SIF do not contain the effects of crack interaction. To include the Kachanov Technique, which takes crack interaction into account, step 5 of the FEAM procedure (Figure 1) would be changed to the following.

- From the tractions calculated in step 2 the average tractions on each crack are found and both are sent as inputs to the Kachanov Technique. SIF are evaluated at each crack tip including the effects of crack interaction.

Traction averages are calculated from the polynomial coefficents prior to the call for the Kachanov technique.

4.3   Removal of Singularity from Numerical Integrations

The accuracy of the FEAM method depends on computational factors such as the method of integration used to evaluate the SIF. Outlined are two of the procedures used in this work.

Coordinate Transformation– The SIF generated by a pair of balanced point forces applied at an arbitrary location on a crack line of length $2l$ are:

$$K_i = \frac{F_i}{(\pi l)^{1/2}} \left\{ \frac{l \pm x}{l \mp x} \right\}^{1/2}$$

(8)

where x is the distance from the crack center to the location of the point force F_i, and F_i for $i = 1, 2$ correspond to the mode I and mode II loadings.

If the distance from the crack tip to the application point of F_i is defined by ξ then

$$K_i = \int_0^{2l} \frac{F_i(\xi)}{(\pi l)^{1/2}} \left\{ \frac{2a - \xi}{\xi} \right\}^{1/2} d\xi \tag{9}$$

If we let $du = d\xi/\xi^{1/2}$ and $\xi = u^2/4$ then

$$K_i = \int_0^{2(2l)^{1/2}} \frac{F_i(u^2/4)}{(\pi l)^{1/2}} [2l - u^2/4]^{1/2} du \tag{10}$$

which is non-singular for $0 \leq u \leq 2(2l)^{1/2}$.

Separate Tip Evaluation– The singularity in the evaluation of the SIF occurs at the crack tips. If the two tips are removed from the evaluation, the integration can be successfully done for the remaining points. The stress at the tips is then evaluated, multiplied by the length parameter coefficient, and added to the result from the modified integration.

5   Results and Discussion

The goal of this work is to establish the feasibility of the Kachanov Method as a part of an effective finite element alternating method to analyze crack interaction problems with finite boundaries. To evaluate the performance of the FEAM it is applied to a series of crack problems found in the literature. In this series results are compared to various sources of both analytical and numerical nature. Following these comparisons is a parametric study of the effects of boundaries on two collinear cracks. The code was run on a Convex mini-supercomputer with even the most complex models taking no more than 4 seconds of cpu time to execute.

All the analyses are considered plane stress with unit thickness, $E = 30000.0$, $\nu = .3$, and loaded by $\sigma_y = 1.0$. Two mesh sizes were investigated, a 48 element coarse grid and a 192 element fine grid. These were used on plate geometries of two different sizes, 6x2 and 2x1. The particular geometries were chosen to match models given in the literature [5]–[7]. Each of the crack problem geometries, the results, and comparisons are presented in the following subsections.

5.1   Single Crack in Mode I Configuration

The purpose of the single crack mode I models is to verify the performance of the FEAM without the Kachanov Technique. All the cases presented in this subsection are center cracked 6x2 plates (Figure 2). Four cases are run for the coarse mesh with crack lengths($2a$) running from 1.6 to 1.0 . As the geometry is symmetric and the loading is pure mode I, a single K_I value is tabulated and compared to results from Isida as referenced by Tada, Paris, and Irwin [5](Table 1). Values from Isida are derived from a Laurent series expansion of the complex stress potential and for all practical purposes are exact.

5.2   Single Inclined Crack

The single crack problem was extended to inclined cracks to examine the ability of the FEAM to handle the inclusion of mode II behavior. The case of one centered crack at a 45° angle in a 2x1 sheet was run for both meshes at a crack length of .4 (Figure 3) .

Results are given in terms of the normalized mode I and mode II SIF (F_I, F_{II}), where $F_I = K_I/\sigma\sqrt{\pi a}$ and $F_{II} = K_{II}/\sigma\sqrt{\pi a}$. Results (Table 2) are compared to FEAM values from Chen[7].

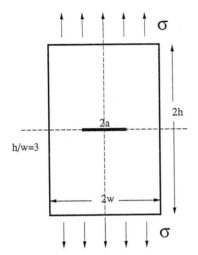
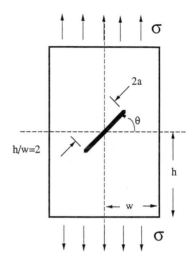

Figure 2: Center Cracked Plate   Figure 3: Plate With Inclined Crack

Case	a	Result	Isida	% Error
1	.8	2.791	2.879	3.1
2	.7	2.146	2.207	2.8
3	.6	1.751	1.789	2.1
4	.5	1.468	1.487	1.3
5	.2	0.811	0.812	0.1

Table 1: Results for Center Cracked Plate

Case	Mesh	a	F_I			F_{II}		
			Result	Chen	% Difference	Result	Chen	% Difference
1	Coarse	.2	.518	.517	0.2	.503	.508	0.9
2	Fine	.2	.521	.517	0.7	.502	.508	1.7

Table 2: Results for Inclined Crack at 45°

5.3 Discussion–Single Crack Models

To this point, single crack FEAM models have been presented. The following is a discussion of errors in the FEAM *without* the Kachanov method. There are two main sources of error.

- Errors due to the numerical evaluation of integrals and the use of the computer for performing numerical calculations.

- Errors due to the approximate nature of the finite element formulation.

The first case of errors is inherent in all computational work. In the FEAM, integrals need to be taken to calculate stresses generated at a point by tractions on a crack line and to calculate SIF at crack tips due to tractions on a crack line. There is additional numerical error in interpolating the stress on the crack lines from the array of elemental stresses. All other computations, including the trigonometric calculations involved in determining the loads at the boundary nodes, also introduce numerical error.

The second type of error, due to the finite element formulation, is mainly dependent on the type of element used. The four noded quad elements are bilinear in nature and can only match linear variations without error. Modeling higher order behavior will result in errors depending on the mesh size and the gradients being approximated. The error in a specific FEAM model is comprised of all the above contributions.

From the single mode I models (Table 1) it can be seen that the inherent error in the non-interacting FEAM is in the range of 0.0% to 3.0% for a realistic range of crack sizes. Errors increase with crack size, and this is explained physically as the bigger the crack, the smaller the ligament between the crack tips and the plate edge. A smaller ligament will exhibit a more rapid stress variation in that region which may require refined modeling. The single inclined case 1 (Table 2) shows a similar difference for mode I and a slightly higher difference for mode II. A curious result is the *increase* in difference in case 2 with the finer mesh. This is explained by considering the two main causes for error. In case 1 it is apparent that the coarse mesh handles the model very well, that is the errors due to the mesh size and the numerics are both small. The stress resolution of the fine mesh is not needed (there may be a *small* reduction in FEM error due to the finer mesh); instead the increased number of integrations and more importantly the increased number of loads that need to be calculated on the boundary (the finer mesh has more boundary nodes) create more numerical error and actually increase the overall error of the model.

5.4 Two Collinear Cracks

The first type of geometry tested with multiple cracks is the case of two collinear cracks symmetrically located in a 6x2 sheet(Figure 4). Three cases were run for both meshes at $2a = .4$ with $2b$, the distance between crack centers, varying from .8 to 1.4 and comparisons are made to Gupta and Erdogan as referenced in Sih[6](Table 3). As in the previous subsection, results are presented as normalized values with alphabetic subscripts identifying the tips, A referring to the inner tips and B to the outer tips. Since the crack length $2a$ is held constant in Table 3, increasing a/b indicates decreasing distance between the inner crack tips.

5.4.1 Discussion–Two Collinear Cracks

With the consideration of two collinear cracks the Kachanov method is now included in the FEAM. From Table 3 it is apparent that errors have increased from the single crack case. This is to be expected as the consideration of two cracks adds to the number of numerical errors of the type found and discussed in the single crack models, and also adds the small error inherent to the Kachanov method.

Case 1 ($a/b = .2857$) is extreme in that the outer crack tip (Tip B) is *very* close to the plate edge (ligament is 5% of total plate width). An expected high difference is seen at the outer

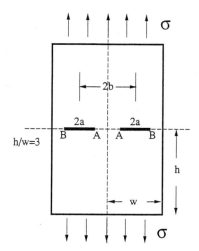

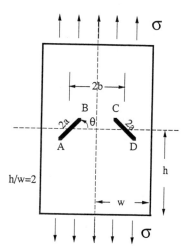

Figure 4: Two Collinear Cracks   Figure 5: Two Inclined Cracks

			F_{IA}			F_{IB}		
Case	Mesh	a/b	Result	Sih	% Difference	Result	Sih	% Difference
1	Coarse	.2857	1.106	1.129	2.0	1.109	1.228	9.6
2	Fine	.2857	1.158	1.129	2.6	1.161	1.228	5.4
3	Coarse	.3333	1.100	1.098	0.2	1.106	1.125	1.7
4	Fine	.3333	1.105	1.098	0.6	1.110	1.125	1.3
5	Coarse	.5000	1.107	1.110	0.3	1.087	1.096	0.8
6	Fine	.5000	1.109	1.110	0.1	1.089	1.096	0.6

Table 3: Results for Plate with Two Collinear Cracks

	Result(Coarse)	Result(Fine)	Chen	% Diff(Coarse)	% Diff(Fine)
F_{IA}	.498	.498	.518	3.8	3.8
F_{IB}	.569	.571	.595	4.2	4.1
F_{IC}	.573	.572	.595	3.8	3.8
F_{ID}	.496	.497	.518	4.3	4.1
F_{IIA}	.529	.529	.552	4.1	4.1
F_{IIB}	.493	.492	.511	3.6	3.7
F_{IIC}	-.493	-.493	-.511	3.6	3.6
F_{IID}	-.529	-.529	-.552	4.1	4.1

Table 4: Results for Plate with Two Cracks Inclined at $\theta = \pm45° : a/b = .91$

tip. The inner tip for case 1, which at first thought would have the lowest difference of the three coarse cases due to its large inter-crack distance, shows sizable difference due to boundary proximity. The remaining two coarse cases show fairly good results with differences remaining, in general, 2% or below. The comparison between the coarse (cases 1,3,5) and the fine (cases 2,4,6) meshes is similar to the behavior found in the single crack cases. For situations where a high level of FEM induced error is expected (Tip B cases 1,2) the benefits of the finer mesh far outweigh the increase in numerical error and the difference is reduced significantly. For example, changing the mesh from coarse to fine for crack geometry $a/b = .2857$, reduces F_{IB} by 44%. In situations with low overall error (Tip A cases 3,4) the finer mesh has a difference that is significantly higher (yet still very acceptable) and in the other cases with moderate levels of error there is only slight fluctuations in difference between the meshes.

5.5 Two Inclined Cracks

The most complicated crack geometry investigated is the case of a sheet with two cracks inclined at $\pm\theta$. For these cases the plate dimensions are 2x1 (Figure 5). A case was run at $\theta = \pm45°$ (Table 4) for both the coarse and fine mesh. $2a$ was held at .5 and $2b$ was held at .55. Results are presented in normalized form and compared to Chen [7]. Once again the %differences are reasonable, generally under 5%. It can be seen again that for models without extreme conditions of crack/crack or crack/boundary proximity, the finer mesh is unnecessary and often does not increase the accuracy of the solution. Based on these cases, the ability of the FEAM/Kachanov technique to solve bounded multiple crack problems has been established.

6 Parametric Study

With the performance and accuracy of the FEAM established by the previous results a study of the effects of boundary proximity is done. The case of two collinear cracks is reconsidered (Figure 4) and cases are run for various a and b. Models with three different crack lengths ($a/w = .1, .2, .25$ where $2w = 2.0$) are investigated with $2b$ varying from 1.6 to .3. Normalized results for tip A (Figure 6) and tip B (Figure 7) are plotted against the ratio of crack length to distance between crack centers (a/b). For reference the case with infinite boundaries [13] is also shown. The normalized result (F_I) is termed the *Correction Factor* as it shows how much the SIF is amplified from the standard SIF for a single crack in an infinite sheet ($\sigma\sqrt{\pi a}$) and represents the effects of boundaries and interaction with other cracks. The coarse mesh was used throughout these cases.

Examining the results for the inner tip, A (Figure 6) the expected result of increasing F_I with increasing a/w is evident. This makes sense physically since as the crack length increases in a finite plate, the influence of the boundary and crack tip interaction will increase. Another physically intuitive result is the trend toward the infinite case as a/b goes to 1.0, *i.e.* as the distance between the inner tips gets small the effects of the boundary are greatly diminished. Another interesting feature of the finite curves, in comparison to the infinite case, is the behavior as a/b decreases. For the infinite case F_{IA} monotonically decreases as a/b decreases. In the case with finite boundaries, as a/b decreases the inner tips move apart and the effects from interaction with the boundaries become significant. Figure 6 shows this boundary effect on F_{IA} at small a/b. For example for cracks $\frac{1}{4}$ the width of the plate, $a/w = .25$, F_{IA} starts to rise for $a/b \leq .5$. The results for the correction factor F_{IB} at the outer crack tip (B) are shown in Figure 7. Although there is less interaction effects at tip B, as compared to tip A (Figure 6), the boundary effects on F_{IB} are evident for smaller a/b.

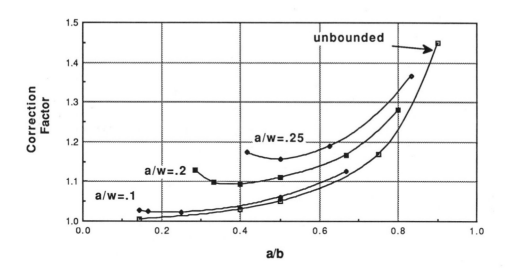

Figure 6: Correction Factor F_{IA} for Inner Crack Tip(A)

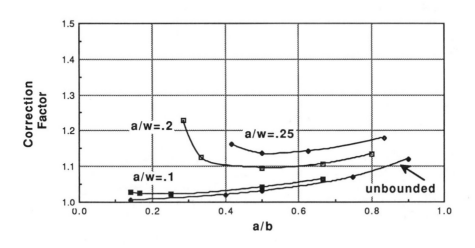

Figure 7: Correction Factor F_{IB} for Outer Crack Tip(B)

7   Conclusions and Future Work

The prevention of failure in structural components relies on design concepts that are damage tolerant. This requires the accurate prediction of important characteristics such as fracture strength and rates of crack growth. For these analyses the stress intensity factors (SIF) associated with the design problem need to be evaluated. For the complicated boundary, constraint, and loading conditions found in structures today the design engineer relies on numerical techniques such as the finite element alternating method (FEAM). The goal of this work is to further advance this method through the introduction of the Kachanov technique. The following conclusions were reached.

- The Kachanov technique can be used efficiently in conjunction with FEAM to generate accurate SIF for bounded multiple crack problems.

- Errors associated with the FEAM depend on both the character of the elements and mesh and numerical factors. Highly discretized meshes, while more computationally intensive, do not necessarily yield better results.

- While not as significant as crack tip interaction, the presence of boundaries increases the SIF for all realistic crack geometries. For bounded crack problems the deviation from "unbounded" behavior must be considered, especially as a/b (the ratio between crack length and intercrack distance) approaches zero and boundary/crack interaction becomes significant.

Of equal importance to the above conclusions is the establishment by this work of the potential of the inclusion of the Kachanov technique in the FEAM. To take full advantage of this potential some improvements to the present work can be made. The following outlines the possible avenues for future improvements.

- Increase the order of the elemental shape functions. Eight-noded or twelve-noded isoparametric elements would be the most likely choice. The use of higher order quads would reduce the number of elements necessary in the mesh and significantly improve the accuracy of the method, especially in the cases of close proximity of tip to tip, or tip to boundary. This would also guarantee good results should more complicated loading conditions be considered.

- Include adaptive meshing capabilities. The present code has standard meshes that do not vary with crack location. While this is simple to implement and is shown to be fairly accurate, often the location of cracks fall in areas of the mesh with the least accurate stress evaluation. Adaptive routines that would consider crack location before meshing would lead to greater accuracy.

- Increase the number of points at which the elemental stresses are evaluated from one to nine(at each of the nine Gaussian points). While a single point is adequate for the presented geometries and load conditions, the nine point evaluation would be necessary for accuracy in models with more complicated boundaries, more complex loads, and greater numbers of cracks.

- The present code is restricted to rectangular geometries. The benefits of the ability to evaluate other more arbitrary geometries is obvious.

- The Kachanov technique has been previously applied to anisotropic problems in unbounded sheets[14]. The theory of this paper can be used to construct a FEAM for the solution of bounded anisotropic problems.

References

[1] S. N. Atluri, S. G. Sampath, and P. Tong. *Structural Integrity of Aging Airplanes.* Springer-Verlas, 1990.

[2] M. Kachanov. Elastic solids with many cracks: A simple method of analysis. *Int. J. Solids Structures*, 23(1):23–43, 1987.

[3] T. S. Moore. Analysis of two-dimensional interacting crack problems using a finite element alternating method. Master's thesis, Tufts University, 1993.

[4] C. Mauge and M. Kachanov. On interaction of cracks in anisotropic solids. In G. Dvorak and D. Lagoudas, editors, *Microcracking Induced Damage in Composites*, pages 95–99. ASME, 1990.

[5] H. Tada, P. Paris, and G. Irwin. *The Stress Analysis of Cracks Handbook.* Del Research Corporation, 1985.

[6] G. C. Sih. *Handbook of Stress Intensity Factors.* Institute of Fracture and Solid Mechanics, Lehigh University, Bethlehem, PA 18015, 1985.

[7] W. Chen and C. Chang. Analysis of two-dimensional mixed mode crack problems by finite element alternating method. *Computers & Structures*, 33(6):1451–1458, 1989.

[8] I. S. Raju and W. B. Fichter. A finite element alternating method for two dimensional mode i crack configurations. *Engineering Fracture Mechanics*, 33(4):525–540, 1989.

[9] S. N. Atluri and T. Nishioka. Analytical solution for embedded elliptical cracks, and finite element alternating method for elliptical surface cracks, subjected to arbitrary loadings. *Engineering Fracture Mechanics*, 17(3):247–268, 1983.

[10] S. N. Atluri, H. L. Simon, and P. E. O'Donoghue. A finite-element-alternating technique for evaluating mixed mode stress intensity factors for part-elliptical surface flaws. *International Journal for Numerical Methods in Engineering*, 24:689–709, 1987.

[11] S. N. Atluri, P. E. O'Donoghue, and T. Nishioka. Analysis of interaction behavior of surface flaws in pressure vessels. *ASME Journal of Pressure Vessel Technology*, 108:24–32, 1986.

[12] O. Orringer, S. French, and M. Weinreich. *User's Guide for the Finite Element Analysis Library (FEABL 2, 4, and 5) and the Element Generator Library (EGL).* Department of Aeronautics and Astronautics, MIT, 1978.

[13] D. P. Rooke and D. J. Cartwright. *Compendium of Stress Intensity Factors.* London: H.M.S.O., 1976.

[14] C. Mauge and M. Kachanov. Interacting arbitrarily oriented cracks in anisotropic matrix. stress intensity factors and effective moduli. *International Journal of Fracture*, 58:R69–R74, 1992.

AD-Vol. 36, Fatigue and Fracture of Aerospace Structural Materials
ASME 1993

CRACK TIP DISPLACEMENT FACTOR, D_I, FOR MODE I CRACKS WITH POWER LAW HARDENING BEHAVIOR

Jiping Zhang
Siemens Power Corporation
Milwaukee, Wisconsin

ABSTRACT

The crack tip displacement analysis is extended to the elastic-plastic fracture mechanics (EPFM) problems for power law hardening materials under the Mode I loading condition. The redefined crack tip displacement factor, D_I, has the form:

$$D_I = \lim_{r \to 0} \left[\frac{COD(r)}{(2r)^{\frac{1}{n+1}}} \right]^{\frac{n+1}{2n}}$$

where $COD(r)$ is the crack opening displacement at a distance, r, from the crack tip, and n is the work-hardening exponent. When $n = 1$, the crack tip displacement factor, D_I, reduces to the format for linear elastic fracture mechanics. For elastic-perfectly plastic materials, ($n = \infty$), the crack tip displacement factor, D_I, equals to the square root of the crack tip opening displacement (**CTOD**). The relations of the **D**-factor with other fracture control parameters, such as **J** and **CTOD** have been given in this article and methods for determination of the **D**-factor are also discussed.

INTRODUCTION

In elastic-plastic fracture mechanics (EPFM), two of the most used fracture control parameters are the J-integral and crack tip opening displacement (CTOD). Due to its straight forward physical meaning, the CTOD method has been widely used in engineering practice since it was introduced by Wells[1,2]. However, there is still no concurrence of opinion as to the significance of the CTOD method[3-7]. At the very crack tip the displacement is actually zero (except for the elastic-perfectly plastic materials) and the CTOD's discussed so far are considered as an effective CTOD which is dependent upon the definition. The most popular definition of CTOD suggested by Rice[8] defined CTOD as the opening distance between the intercept of two 45°-lines, drawn back from the tip with the deformed profile, as shown in Figure 1. This definition can be applied to Mode I or Mode III cracks but definitely not to Mode II cracks, nor mixed mode cracks. Although CTOD can be related to the linear elastic fracture mechanics (LEFM) parameters through small scale yielding assumption, it is virtually not applicable to

LEFM problems.

On the other hand, with the advance of the numerical tools such as finite element method (FEM) and boundary element method (BEM), the crack surface displacements are easy to obtain and are used to evaluate the fracture behavior of a structure. The crack surface displacements were widely used to get stress intensity factors in LEFM problems by many investigators. However, the accuracy of the results depend upon how valid is the assumption of the square root of r behavior of the crack tip displacement since the nodal points are away from the crack tip. For 2D analysis, the crack tip elements can be small enough to obtain the required accuracy, but a fine mesh with small crack tip element for a 3D analysis is very difficult to achieve due to the limit of the element aspect ratio. Based on some previous studies on the crack surface relative displacements[9-12] for mixed mode fracture problems, Zhang and Ghorbanpoor introduced a new fracture control parameter, D, which utilizes the crack tip displacements normalized by the square root displacement behavior [13-15]. The proposed crack tip displacement factors, D, showed very promising potentials especially for three-dimensional LEFM analysis using numerical tools.

This paper demonstrates how the crack tip displacement factor, D, is extended to Mode I EPFM problems for power law hardening materials where the HRR singularities are dominated at the crack tip. A finite element analysis was performed to illustrate how to obtain the D-factor from the numerical results. The relations between the crack tip displacement factor and other fracture control parameters are also discussed.

BACKGROUND INFORMATION

HRR Singularities:

Consider a non-linear elastic-plastic material where the stress-strain curve can be represented by the Ramberg-Osgood equation:

$$\frac{\varepsilon}{\varepsilon_o} = \frac{\sigma}{\sigma_o} + \alpha(\frac{\sigma}{\sigma_o})^n \qquad (1)$$

where σ_o is the yield stress, ε_o is the strain at yield, n is the work-hardening exponent and α is material constant. If the linear elastic strain is small with respect to the non-linear portion, the above equation can be simplified to:

$$\frac{\varepsilon}{\varepsilon_o} = \alpha(\frac{\sigma}{\sigma_o})^n \qquad (2)$$

According to Hutchingson[16], Rice and Rosengren[17], the stress and strain fields (referred to as HRR-singularities) are as follows:

$$\sigma_{ij} = (\frac{J}{\alpha \sigma_o \varepsilon_o I_n r})^{\frac{1}{n+1}} \tilde{\sigma}_{ij}(\theta) \qquad (3)$$

$$\varepsilon_{ij} = \alpha \varepsilon_o (\frac{J}{\alpha \sigma_o \varepsilon_o I_n r})^{\frac{n}{n+1}} \tilde{\varepsilon}_{ij}(\theta) \qquad (4)$$

126

and the corresponding displacement field is:

$$u_i = \alpha \varepsilon_o (\frac{J}{\alpha \sigma_o \varepsilon_o I_n})^{\frac{n}{n+1}} r^{\frac{1}{n+1}} \tilde{u}_i(\theta) \qquad (5)$$

where J is the J-integral and I_n is a numerical constant where the value depends upon the stress-strain relation.

Crack Tip Displacement Factor in LEFM Problems:

The concept of crack tip displacement factors was proposed initially for linear elastic fracture mechanics (LEFM) analysis [13-15]. The D-factors can be expressed in the following forms:

$$D_I = \lim_{r \to 0} \frac{COD(r)}{F(r)} \qquad (6)$$

$$D_{II} = \lim_{r \to 0} \frac{CSD(r)}{F(r)} \qquad (7)$$

$$D_{III} = \lim_{r \to 0} \frac{CTD(r)}{F(r)} \qquad (8)$$

where COD(r), CSD(r) and CTD(r) are crack surface relative displacements for Mode I, II, III cracks, respectively, r is the distance from the crack tip, and F(r) is a normalizing function which approaches square root of 2 times the distance r when r approaches zero as follows:

$$\lim_{r \to 0} F(r) = \lim_{r \to 0} \sqrt{2r} \qquad (9)$$

and the D-factors are related to stress intensity factors by:

$$K_I = \frac{E^* \sqrt{\pi}}{4} D_I \qquad (10)$$

$$K_{II} = \frac{E^* \sqrt{\pi}}{4} D_{II} \qquad (11)$$

$$K_{III} = \alpha \frac{E^* \sqrt{\pi}}{4} D_{III} \qquad (12)$$

where $E^* = E$ for plane stress conditions and $E^* = E/(1 - \nu^2)$ for plane strain conditions, and $\alpha = 1 + \nu$ is the constant for Mode III cracks.

The total crack tip displacement, **D**, can be written in the form of:

$$D^2 = D_I^2 + D_{II}^2 + D_{III}^2 = \frac{16G}{\pi E^*} \qquad (13)$$

or expressed as:

$$D = \lim_{r \to 0} \frac{CSRD(r)}{F(r)} = 4\sqrt{\frac{G}{\pi E^*}} \quad (14)$$

where **CSRD(r)** is the crack surface relative displacement at the sampling location and **G** is the total energy release rate.

The above equations show that the D-factors are fully convertible to other fracture parameters in LEFM. The D-factors can be obtained by analytical approach if an expression of the crack surface displacement is given, or by extrapolating techniques for numerical analysis and experimental results. Examples of the application to two- and three-dimensional LEFM can be found in the author's previous work[14-16].

D-FACTOR FOR EPFM PROBLEMS

The D-factor defined by Equations (6) to (8) are limited to the linear elastic fracture mechanics problems. For a crack tip displacement field governed by the HRR singularities, the crack tip displacement factor, D_I, in Mode I cracks is redefined by the following equation:

$$D_I = \lim_{r \to 0} \left[\frac{COD(r)}{(2r)^{\frac{1}{n+1}}}\right]^{\frac{n+1}{2n}} \quad (15)$$

The **D**-factor agrees with the format used in the LEFM problems when **n** is set to 1, as shown in Equation (6). For elastic-perfectly plastic materials, $n \to \infty$, the crack tip displacement factors can be rewritten as:

$$D_I = \lim_{r \to 0} \sqrt{COD(r)} = \sqrt{CTOD} \quad (16)$$

Please note here the **CTOD** is defined as the crack tip opening displacement for ideal elastic-perfectly plastic materials only, of which the **CTOD** value is not zero. From Equation (5), we know that **COD(r)** equals:

$$COD(r) = 2\alpha\epsilon_o \left(\frac{J}{\alpha\epsilon_o\sigma_o I_n}\right)^{\frac{n}{n+1}} r^{\frac{1}{n+1}} \tilde{u}_2(\pi) \quad (17)$$

and the corresponding crack tip displacement factor is:

$$D_I = \lim_{r \to 0} \left[\frac{COD(r)}{(2r)^{\frac{1}{n+1}}}\right]^{\frac{n+1}{2n}}$$

$$= \left[\frac{2\alpha\epsilon_o \left(\frac{J}{\alpha\epsilon_o\sigma_o I_n}\right)^{\frac{n}{n+1}} \tilde{u}_2(\pi)}{2^{\frac{1}{n+1}}}\right]^{\frac{n+1}{2n}}$$

$$= [2(\alpha\epsilon_o)^{\frac{1}{n}}\bar{u}_2(\pi)^{\frac{n+1}{n}}/I_n \frac{J}{\sigma_o}]^{\frac{1}{2}}$$

$$= [d_n^* \frac{J}{\sigma_o}]^{\frac{1}{2}} \qquad (18)$$

where

$$d_n^* = 2(\alpha\epsilon_o)^{\frac{1}{n}}\bar{u}_2(\pi)^{\frac{n+1}{n}}/I_n$$

Equation (18) relates the crack tip displacement factor, D_I, to the J-integral. The reverse form of the above equation is:

$$J = \sigma_o D_I^2/d_n^* \qquad (19)$$

The crack tip displacement factor, D_I, can be also related to the conventional CTOD parameter, δ_{45}, by the following equation:

$$D_I = [\frac{d_n^*}{d_n} \delta_{45}]^{\frac{1}{2}} \qquad (20)$$

or

$$\delta_{45} = \frac{d_n}{d_n^*} D_I^2 \qquad (21)$$

where d_n is a constant given by Shih[18] which has the form:

$$d_n = 2(\alpha\epsilon_o)^{\frac{1}{n}}[\bar{u}_1(\pi) + \bar{u}_2(\pi)]^{\frac{1}{n}}\bar{u}_2(\pi)/I_n$$

The dependence of d_n^* upon n and ϵ_o is very similar to that of d_n. For a perfectly plastic behavior ($n \rightarrow \infty$) under plane stress, both d_n and d_n^* are equal to one, Equation (20) coincides with Equation (16).

Equations (18) to (21) establishes an equivalence between D_I and J, D_I and δ_{45}. Therefore, any fracture criterion based upon either a critical value of J or a critical value of δ_{45} is equivalent to one based upon a critical value of the crack tip displacement factor, D_I and vice versa. In order to ensure the validity of Equations (18) and (19), the HRR field must engulf the crack tip. In addition, the definition of δ_{45} requires that the HRR field must dominate the crack tip deformation zone over a region at least as large as δ_{45}. Although the current study is given to Mode I cracks in this article, the methodology can be extended to the other two modes.

DETERMINATION OF CRACK TIP DISPLACEMENT FACTOR

If there exists an expression for the crack surface displacement as a function of r, a closed form solution is possible by using the approach similar to that for LEFM problems. However, due to the complexity of materials non-linearity, such a closed form solution is virtually impossible. Therefore, one has to rely on alternative ways to derive the crack tip displacement factor, such as numerical or experimental methods. If solutions have been obtained from J-integral method or CTOD method (δ_{45}), one can derive the D-factor using Equations (18) or (20) because of their equivalence. For most engineering problems needing to be solved, numerical analyses, such as FEM and BEM, may be the best approaches, for the crack surface displacements are already in their solutions.

A center cracked plate (half crack size, a = 1; half plate width, W = 100; half plate height, H = 200) was used to illustrated how to determine the D-factor using the results from numerical analysis. The material used in the example was 304 stainless steel and the material properties were:

> Elastic Modules, E = 29.4E6 psi
> Yield Stress, σ_o = 38,400 psi
> Constant, α = 5.682E-3
> Work Hardening Exponent, n = .5

A finite element analysis was performed for the above example and the FEA model is shown in Figure 2. The crack surface displacement results were used to calculate the normalized **D(r)** values. For LEFM analysis, the crack tip displacement factor, **D**, can be obtained by extrapolating the data points (**D(r)**) to the crack tip in the regular linear coordinate system. However, for EPFM problems, an extrapolation to the crack tip is not necessary. From Equation (18), we know that inside the HRR deformation zone, the D-factor is independent of the sampling distance, **r**. Therefore, the D-factor should be constant within the plastic zone. However, it is suggested that the nodal point next to the crack tip be used to calculate the D-factor if there are several nodal points inside the plastic zone. If there is no nodal point inside the plastic zone, the D-factor can be estimated by extrapolating the D(r) curve to the elastic-plastic boundary close to the crack tip. Such an extrapolation should be performed in a logarithmic coordinate system. The D-factor can be obtained by extrapolating the **D(r)** values to the boundary of plastic zone.

Figure 3 shows the D(r) curves in a log scale for different applied load levels. It is seen that at higher load levels the **D(r)** curves become flat for those nodal points close to the crack tip, which implies they are inside of the plastic zone. The J-integral values were collected from the same FEA model using a path of the innermost circle. The values of $\sigma_o D_i^2/J$ (d_n^*) were calculated for various load levels as illustrated in Figure 4. The d_n^* curve is almost constant for the higher load levels. The slightly lower values at lower load levels were due to the error from J-integral computation since the plastic zone didn't engulf the complete integral path at lower stress magnitude.

CONCLUSION

The crack tip displacement analysis was extended to materials with power law hardening behavior where the HRR singularities dominate the crack tip. The redefined crack tip displacement factor, D, can also be used for elastic-perfectly plastic materials in addition to the regular LEFM problems. Due to its equivalence between D-factor, J-integral and CTOD (δ_{45}), the crack tip displacement factor, D, can be used as a fracture criterion. The determination of the D-factor is very straight forward once the crack surface displacements are known. Although the current study in this article is for Mode I cracks, similar methodology can be applied to the Mode II and Mode III problems.

130

REFERENCES

[1]   Wells, A. A., "Unstable Crack Propagation in Metals-Cleavage and Fast Fracture", Proceedings Crack Propagation Symposium, Cranfield, pp. 210-230, (1961).

[2]   Wells, A. A., "Application of Fracture Mechanics At and Beyond General Yielding", British Welding Research Ass., Rep. M13, (1963).

[3]   Srawley, J. E., Swedlow, J. L., and Roberts, E., "On The Sharpness of Cracks Compared With Wells' COD Method", International Journal of Fracture Mechanics, Vol. 6, pp. 441-444, (1970).

[4]   Wells, A. A. and Burdekin, F. M., Discussion to [3], International Journal of Fracture Mechanics, Vol. 7, pp. 233-241, (1971).

[5]   Srawley, J. E., Swedlow, J. L. and Roberts, E., Response to [4], International Journal of Fracture Mechanics, Vol. 7, pp. 242-246, (1971).

[6]   Wells, A. A., "Crack Opening Displacements From Elastic-Plastic Analysis of Externally Notched Bars", Engineering Fracture Mechanics, Vol. 1, pp. 399-410, (1969).

[7]   Elliott, D., Walker, E. F. and May, M. J., "The Determination and Applicability of COD Test Data", Conference - Practical Applications of Fracture Mechanics to Pressure Vessel Technology (1971)

[8]   Tracy, D. M., "Finite Element Solutions for Crack-Tip Behavior in Small-Scale Yielding", Journal of Engineering Material Technology, Vol. 98, pp. 146-151, (1976).

[9]   Feng, H., Zhang, J., Rohde, J. and Ghorbanpoor A., "Analysis of Crack Surface Relative Displacement for Mixed Mode I and Mode II Cracks," Proceedings in The 21st Mid-Western Mechanics Conference, Houghton, MI, Aug. 14-16, (1989).

[10]   Zhang, J. and Ghorbanpoor, A., "CTRD Method for SSY Mixed Mode Fracture Problems", Proceedings in The International Symposium Testing & Failure Analysis, Los Angeles, CA, Nov., (1989).

[11]   Ghorbanpoor, A. and Zhang, J., "Boundary Element Analysis of Crack Growth for Mixed Mode Center Slant Crack Problems", Engineering Fracture Mechanics V36, n5, pp. 661-668, (1990).

[12]   Feng H., Zhang, J. and Rohde, J., "The Crack Surface Relative Displacement of Mixed Mode Fracture Problems", Engineering Fracture Mechanics, V36, n6, pp. 971-978, (1992).

[13]   Zhang, J., "Boundary Element Analysis of Fracture Mechanics Problems", Ph.D. Dissertation, University Wisconsin-Milwaukee, (1991).

[14]   Zhang, J. and Ghorbanpoor, A., "Crack Tip Displacement Factor-D: An Application to Two Dimensional Linear Elastic Fracture Mechanics", International Journal of Fracture, Vol. 59, pp. 133-150, (1993).

[15]   Zhang, J. and Ghorbanpoor, A., "Crack Tip Displacement Factor-D: An Application to Three-Dimensional Linear Elastic Fracture Mechanics," ready for publication (1993).

[16]   Hutchinson, J. W., "Singular Behavior at the End of a Tensile Crack in a Hardening Material", Journal Mechanical Physical Solids, Vol. 16, pp. 13-31, (1968).

[17]   Rice, J. R. and Rosengren, G. F., "Plane Strain Deformation Near a Crack Tip in a Power-Law Hardening Material", Journal of Mechanics and Physical Solids, Vol. 16, pp. 1-12, (1968).

[18]   Shih, C.F., "Relationship Between the J-Integral and the Crack Opening Displacement for Stationary and Extending Cracks," Journal of the Mechanics and Physics od Solids, Vol. 29, pp. 305-326, (1981).

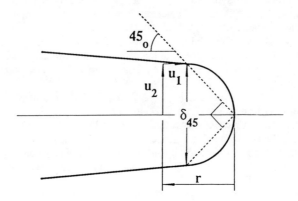

Figure 1.   Conventional Definition of the Crack Tip Opening Displacement (CTOD).

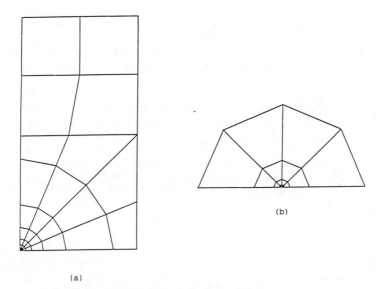

(a)

Figure 2.   Finite Element Mesh of a Center Cracked Plate. (a = 1", W = 100" and H = 200")

(a). One Quarter of The Plate;

(b). Details at the Crack Tip.

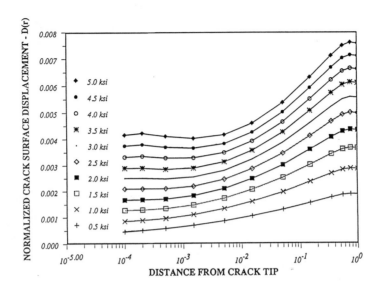

Figure 3.   Normalized Crack Surface Displacements D(r).

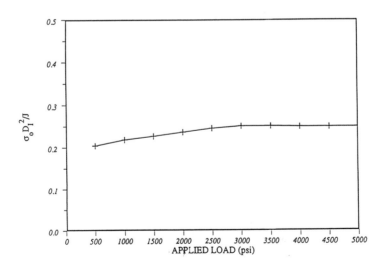

Figure 4. $\sigma_o D_I^2/J$ (d_n^*) Values at Various Load Levels.

AD-Vol. 36, Fatigue and Fracture of Aerospace Structural Materials
ASME 1993

EXPERIMENTAL AND BOUNDARY ELEMENT ANALYSIS
OF HERTZIAN CONE CRACKING

Shi-yew Chen and Thomas N. Farris
School of Aeronautics and Astronautics

S. Chandrasekar
School of Industrial Engineering

Purdue University
West Lafayette, Indiana

ABSTRACT

Quasi-static indentation of brittle materials with a spherical indenter induces Hertzian cone cracks. The variation of cone crack length with indenter load has been investigated experimentally by indenting soda-lime glass with 3.17 mm and 6.34 mm diameter steel balls, respectively. Axisymmetric boundary elements have been used to accurately and efficiently calculate the stress intensity factors along the edge of the cone crack by adapting the modified crack closure integral. The boundary element results have been verified through comparison to finite element results. Based on the assumption that cone crack propagation is arrested when the stress intensity factor at the crack tip is less than the fracture toughness, we have observed reasonably good agreement between experiment and theory. A proposed continual loading scheme is used to establish the boundary condition in Hertzian cone cracking. The cone crack length-load relationship is not sensitive to the ball indenter size within the cone crack loading range and the change in the predicted cone crack length for 25% of variation in the assumed ring crack radius is less than 5%.

INTRODUCTION

The indentation fracture of brittle materials generated by spherical and pointed (i.e. cone, Vickers, Knoop) indenters has been the focus of much research. Interest in this problem stems from the role of indentation-induced flaws in controlling the strength of brittle solids and from the use of indentation techniques to measure fracture surface energy and fracture toughness of brittle solids. In addition there is recent evidence that indentation fracture can be used to model material removal during finishing of ceramics. The indentation by spheres is commonly referred to as Hertzian indentation after Hertz who originally calculated the relevant pressure distribution under the indenter. The Hertzian indentation of brittle materials and the cone cracks that it produces are the focus of this paper (see Figure 1 for a schematic).

Frank and Lawn (1967) investigated the development of the cone crack from the ring crack in the strongly inhomogeneous Hertz stress field. They approximately calculated the Mode I stress intensity factor at the cone crack front by modeling it as a two dimensional plane strain crack in an infinite medium loaded by the stresses acting on the location of the cone crack in the uncracked body. It was assumed that the cone cracked followed trajectories of the minimum principal stress so that it was perpendicular to the maximum principal stress. This model predicts that the crack propagation arrests at a depth related to the applied load.

Warren (1978) performed a similar calculation using a similar two dimensional approximation and considered the effect of the ring crack initiating some distance outside the circle of contact. He combined his model with experiments to measure the fracture toughness of carbides based on the initiation of the ring crack and concluded that the reliability of interpretation can be improved by more sophisticated fracture mechanics analysis. Mouginot and Maugis (1985) again use this approach to

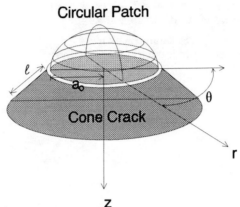

Figure 1 Hertzian cone crack with semielliptical contact pressure distribution

analyze cracks generated by both spherical and flat punches. Again Zeng et al (1992a, b) combine a similar approach with measurements to obtain the fracture toughness of soda lime glass and obtain values of critical stress intensity, K_{Ic}, that are load independent and reasonably consistent with values for K_{Ic} determined by other means. None of the above analyses include the effect of the actual crack shape and its interelation with the free surface on the stress intensity factors.

The mixed mode stress intensity factors for the Hertzian cone crack have been fully analyzed numerically only recently by Li and Hills (1991). A global-local finite element method was used and effects of the free surface and the curved nature of the crack front was included in the analysis. Li and Hills (1991) found that the Mode II stress intensity factor was about one tenth of the Mode I stress intensity factor. It should be noted that their calculation is based on one constant ratio of contact radius to ring crack radius. i.e. $a/a_o = 1.05$. They did not provide a detailed comparison of the calculation with experiments nor were any alternative solutions used to validate the finite element results.

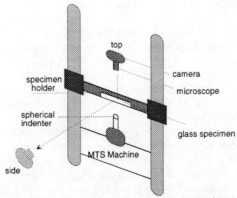

Figure 2 Schematic of the setup for in situ observation of indentation cracking

Table 1 Material properties of soda-lime glass and steel

	Young's Modulus (GPa)	Poisson's Ratio	Fracture Toughness $\left(MPa\sqrt{m}\right)$
soda lime glass	69	0.23	0.75
steel	200	0.25	50-154

In the present study, soda-lime glass blocks are indented with two spherical steel indenters of different sizes. The variation of the cone crack length with applied load in soda-lime glass is determined experimentally from measurements. The angle which the sides of the cone crack makes with the free surface is also measured. An axisymmetric Boundary Element Method (BEM) is developed and applied to calculate the stress intensity factors around the cone crack using the modified crack closure integral (Chen and Farris, 1993). A detailed comparison of the BEM results with the FEM results of Li and Hills (1991) is carried out. Finally, the predicted variation of cone crack length with applied load is compared with the experimental measurements.

EXPERIMENTS

Soda-lime glass specimens were indented by steel spheres. Table 1 gives some important mechanical properties. The soda-lime glass blocks were 5cm x 5cm x 1.25cm (2in x 2in x 0.5in). The blocks were polished to an optical surface quality and indented with hardened steel balls of diameter 3.17mm and 6.34mm respectively. Figure 2 shows a schematic of the indentation set-up. In a typical indentation experiment, the indenter, which was attached to the cross-head of an MTS testing machine (MTS model 458.20), was loaded against the block. The cross-head speed when indenting the glass block was 0.1mm/min. During loading and unloading, the indenter and the block were observed and photographed through the top and side faces, see Figure 2. Particular attention was paid to observing the evolution of the cone crack as the indenter load was increased. From photographs of the cone crack and measurements of the indenter load, the variation of cone crack length with applied load was determined for both the 3.17mm indenter (Figure 3) and the 6.34mm indenter (Figure 4).

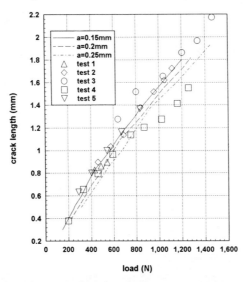

Figure 3 Experimental and predicted cone crack length for 3.17mm (0.125 in) ball indentation on glass using $K_{Ic} = 0.75 MPa\sqrt{m}$, $\alpha=62°$

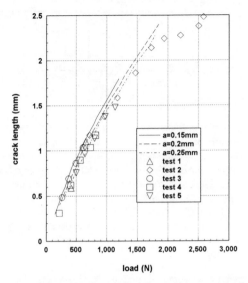

Figure 4 Experimental and predicted cone crack length for 6.34 mm (0.25 in) ball indentation on glass using $K_{Ic} = 0.75 MPa\sqrt{m}$, α=62°

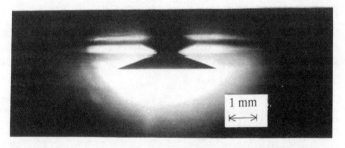

Figure 5 Photo of side view of cone crack on soda-lime glass, 1696 N load, 3.17mm ball indentation

Figure 6 Photo of top view of cone crack on soda-lime glass, 1690N maximum load , 3.17mm ball indentation

The relationship between load and cone crack length is similar for each ball size.

Figure 5 shows a photograph of the side view of the cone crack taken in-situ through the side of the glass block. The load corresponding to the figure was 1696 N. There was stable extension of the cone crack with increasing load. Figure 3 and Figure 4 shows the variation of the cone-crack length with load. Measurements of the angle of intersection of the sides of the cone crack with the surface showed that the half apex angle of the cone varied from 59° to 69° in 10 indentation tests. The top view of the cone crack after indentation is shown in Figure 6. In some of the indentation tests, a median crack, which was deeper than the lower edge of the cone crack, was observed during unloading(Figure 7).

Figure 7 Photo of side view of cone crack and median crack, 715 N load, 3.17mm ball indentation

BOUNDARY ELEMENT ANALYSIS

Brief Development of Axisymmetric BEM

Somigliana's Identity. The boundary integral equation method is based upon the transformation of the partial differential equations to integral equations applicable over the boundary. Details of the following brief development can be found in Kermanidis(1975), Mayr(1976), Cruse et al(1977), and Bakr(1983). According to Betti's reciprocal work theorem, the Somigliana identity can be used to find displacements at an interior point p due to boundary tractions and displacements in the absence of body forces as

$$u_i(p) = -\int_\Gamma t_{ij}^*(p,q)u_j(q)ds(q) + \int_\Gamma u_{ij}^*(p,q)t_j(q)ds(q) \tag{1}$$

where $u_i(p)$ is the displacement in the i-direction at point p, $t_{ij}^*(p,q)$ and $u_{ij}^*(p,q)$ are the traction and displacement at point q in the j direction due to a point load in the i-direction at point p in an infinite linear elastic body, respectively, $u_j(q)$ and $t_j(q)$ are the boundary values of the displacement and traction, and Γ is the surface of the bounded domain Ω. The usual summation convention for repeated indices is used throughout. Equation (1) is known as Somigliana's identity for displacements(Brebbia et al, 1984). The fundamental solutions u_{ij}^* are the solution to Navier's equation for an applied point force satisfying

$$\mu u_{ji,kk}^* + \frac{\mu}{1-2\nu}u_{jk,ki}^* + b_i^* = 0 \tag{2}$$

where μ and ν are the shear modulus and Poisson's ratio, respectively, and

$$b_i^* = \delta_{ij}\Delta(p,q)F_j \tag{3}$$

where

$$\Delta(p,q) = \begin{cases} \infty & p = q \\ 0 & p \neq q \end{cases}$$

(4)

is the Dirac delta function, p and q are vectors designating the load point and the field point, respectively, and F_j represents a point force of magnitude unity in the j direction.

If the load point is on the boundary, the first integral on the right hand side of equation (1) becomes singular. This leads to the following tensorial coefficient defined by the limit of that singular integral

$$c_{ij}(p) = \delta_{ij} + \lim_{\varepsilon \to 0} \int_{\Gamma^\varepsilon} t_{ij}^*(p,q)d\Gamma$$

(5)

where Γ^ε designates a portion of a spherical surface of radius ε centered at the load point p on the surface, as shown in Figure 8.

Assuming the elastic domain is an axisymmetric solid of revolution, the surface integral in Equation (1) can be reduced to a line integral leading to

$$c_{ij}(\xi)u_j(\xi) + 2\pi \int_{\Gamma} \hat{t}_{ij}(\xi,x)u_j(x)r(x)d\Gamma(x) =$$

$$2\pi \int_{\Gamma} \hat{u}_{ij}(\xi,x)t_j(x)r(x)d\Gamma(x),$$

$$i, j = r, z$$

(6)

where the factor $2\pi r(x)$ is the variable radius of the solid of revolution(Figure 9). Equation (6) is valid for interior points($\xi \in \Omega$) with $c_{ij}(\xi) = \delta_{ij}$ and for smooth boundaries where $c_{ij} = 1/2\delta_{ij}$ with δ_{ij} being the

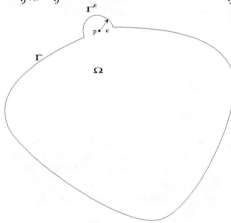

Figure 8 Integration at a singular point p on
the boundary surrounded by part of a
spherical surface with radius of ε

kronecker delta function =1 for i=j and 0 for i ≠ j, and $\hat{t}_{ij}$ and $\hat{u}_{ij}$ are the fundamental solution for a ring load having radius R(ξ) acting in an infinite solid.

Axisymmetric Fundamental Solution. The fundamental solution to equation (2) is known as the Kelvin solution which can be expressed as (Brebbia, 1980)

$$u_{ij}^*(p,q) = \frac{1}{16\pi(1-v)\mu\rho}\{(3-4v)\delta_{ij} + \rho_{,i}\rho_{,j}\}$$

(7)

for three-dimensional problems where ρ is the distance between the load point and the field point.

140

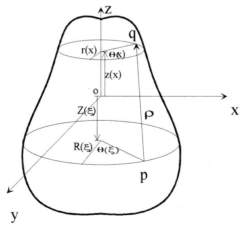

Figure 9 Definition of cylindrical coordinate system

Integrating the three dimensional fundamental solution of Equation (7) along a circle in the cylindrical coordinate system(r,θ,z) (Figure 9) leads to the axisymmetric fundamental solution for ring loads (Figure 10) given by

$$\hat{u}_{rr} = \frac{1}{16\pi(1-v)\mu\sqrt{Rr}}\{(3-4v)Q_{+\,1/2}(\gamma)+\frac{\overline{Z}}{Rr}\frac{dQ_{+\,1/2}}{d\gamma}\},$$

$$\hat{u}_{rz} = \frac{\overline{Z}}{16\pi(1-v)\mu\sqrt{Rr}\,r}\{\frac{Q_{+\,1/2}(\gamma)}{2}-(\gamma-\frac{r}{R})\frac{dQ_{+\,1/2}}{d\gamma}\},$$

$$\hat{u}_{zr} = \frac{-\overline{Z}}{16\pi(1-v)\mu\sqrt{Rr}\,r}\{\frac{Q_{-\,1/2}(\gamma)}{2}+(\gamma-\frac{r}{R})\frac{dQ_{-\,1/2}}{d\gamma}\}, \tag{8}$$

$$\hat{u}_{zz} = \frac{1}{16\pi(1-v)\mu\sqrt{Rr}}\{(3-4v)Q_{-\,1/2}(\gamma)-\frac{\overline{Z}}{Rr}\frac{dQ_{-\,1/2}}{d\gamma}\},$$

where $Q_{+1/2}$ and $Q_{-1/2}$ are Legendre functions of order zero and the following notation is used:

$$Q_{+\,1/2} \equiv Q_{+\,1/2}(\gamma), \quad Q_{-\,1/2} \equiv Q_{-\,1/2}(\gamma)$$

$$\begin{aligned} R &\equiv R(\xi), & z &\equiv z(x), \\ Z &\equiv Z(\xi), & \overline{Z} &\equiv Z-z, \\ r &\equiv r(x), & \gamma &\equiv 1+\frac{\overline{Z}^2+(R-r)^2}{2Rr}, \end{aligned} \tag{9}$$

The Legendre functions, $Q_{+1/2}$ and $Q_{-1/2}$, and their derivatives can be written as (Abramowitz et al, 1972)

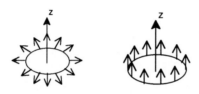

Figure 10 Ring load for radial, and axial directions

$$Q_{+1/2}(\gamma) = \gamma k K(k) - 2E(k)/k,$$

$$Q'_{+1/2}(\gamma) = \frac{k}{2}\{K(k) - \frac{\gamma}{\gamma-1}E(k)\},$$

$$Q_{-1/2}(\gamma) = kK(k),$$

$$Q'_{-1/2}(\gamma) = -\frac{k}{2}\frac{1}{\gamma-1}E(k), \tag{10}$$

in which $K(k)$ and $E(k)$ are the complete elliptic integrals of the first and second kinds, respectively, and the modulus of the elliptic integral is given by

$$k = \sqrt{\frac{2}{1+\gamma}}, \tag{11}$$

The expressions for the corresponding stresses can be obtained from the stress-displacement relations in cylindrical coordinates which can be written as

$$\hat{\sigma}^j_{rr} = \frac{2\mu v}{1-2v}[\frac{\hat{u}_{ir}}{r} + \frac{\partial \hat{u}_{iz}}{\partial z} + (\frac{1-v}{v})\frac{\partial \hat{u}_{ir}}{\partial r}],$$

$$\hat{\sigma}^j_{rz} = \mu(\frac{\partial \hat{u}_{iz}}{\partial r} + \frac{\partial \hat{u}_{ir}}{\partial z}),$$

$$\hat{\sigma}^j_{zz} = \frac{2\mu v}{1-2v}[\frac{\hat{u}_{ir}}{r} + \frac{\partial \hat{u}_{ir}}{\partial r} + (\frac{1-v}{v})\frac{\partial \hat{u}_{iz}}{\partial z}], \tag{12}$$

The superscripts on stress denote the direction in which the load is applied. The tractions required in Equation (6) are evaluated using

$$\hat{t}_{ij} = \hat{\sigma}^i_{kj}n_k, \quad i,j,k = r,z \tag{13}$$

where $\hat{\sigma}^i_{kj}$ is the stress σ_{kj} due to the ring load acting in the i direction and n_k is the outward normal vector. In order to carry out the differentiation of the fundamental displacements, the following relations (Erdelyi, 1953) are useful

$$K'(k) = E(k)/[k(1-k^2)] - K(k)/k,$$

$$E'(k) = [E(k) - K(k)]/k \tag{14}$$

The recursive relation in Equation (14) is used to remove higher order derivatives of the elliptic functions and Legendre equations occurring in the traction kernels.

Boundary Element and Numerical Scheme. Discretizing the boundary Γ into N quadratic boundary elements Γ_m, Equation (6) becomes

$$c_{ij}(\xi)u_j(\xi) + 2\pi\sum_{m=1}^{N}u_{jmk}\int_{-1}^{+1}\hat{t}^*_{ij}(\xi,x)N_k r_m J^m(\zeta)d\zeta = 2\pi\sum_{m=1}^{N}t_{jmk}\int_{-1}^{+1}u^*_{ij}(\xi,x)N_k r_m J^m(\zeta)d\zeta,$$

$$i,j = r,z,$$

$$k = 1,2,3,$$

$$m = 1,2,...,N, \tag{15}$$

where u_{jmk} and t_{jmk} are the nodal displacement and traction in the j-direction of the k-th node of element m, r_m is the radius of any point in the boundary Γ_m, and N_k is the shape function, $J_m(\zeta)$ is the Jacobian of the geometric transformation from the physical plane (r,z) to the ζ-line for element Γ_m, defined by

$$J^m(\zeta) = \begin{vmatrix} e_r & e_z \\ \frac{\partial r_m}{\partial \zeta} & \frac{\partial z_m}{\partial \zeta} \end{vmatrix} \tag{16}$$

142

The shape functions used for quadratic elements are

$$r_m = \sum_{k=1}^{3} r_{mk} N_k, \quad z_m = \sum_{k=1}^{3} z_{mk} N_k$$

$$u_{jm} = \sum_{k=1}^{3} u_{jmk} N_k,$$

$$t_{jm} = \sum_{k=1}^{3} t_{jmk} N_k, \tag{17}$$

$$N_1 = -\zeta(1-\zeta)/2,$$

$$N_2 = (1+\zeta)(1-\zeta),$$

$$N_3 = +\zeta(1+\zeta)/2, \tag{18}$$

where u_{jm} and t_{jm} are the displacement and traction, respectively, in the j direction of element m and ζ is the local element coordinate equal to -1 at the first node, 0 at the second node, and +1 at the third node. Note that integration along the z-axis is not required for a solid model because it is within the interior of the boundary.

By assigning ξ to each node one at a time, 2M equations can be obtained from Equation (15), where M($=2N+1$) is the number of total nodes. The integration of Equation (15) are carried out by gaussian quadrature for the non-singular elements(ξ not on the same element as x) and by logarithmic gaussian quadrature for singular elements on the right hand side of the equation (ξ on the same element as x). The singular term on the left hand side of the equation is evaluated by applying a rigid body displacement in the z-direction and a hydrostatic pressure solution (Bakr, 1985). Prescribing the given boundary conditions leads to 2M simultaneous equations for the 2M unknown boundary tractions and displacements which are solved numerically. Once all of the boundary values are known, stresses and displacements at internal points are found as a summation similar to Equation (15).

Multidomain technique. Consider two domains partially connected together (see Figure 11). Domain Ω_1 and domain Ω_2 share the same boundary Γ_{12}. The two domains may have similar or dissimilar elastic properties. The unshared boundaries of Ω_1 and Ω_2 are Γ_1 and Γ_2 respectively, and u^i and t^i are the displacement and traction for domain i. Considering each domain separately, the BEM equations can be written as

$$\begin{bmatrix} A_1 & A_{12} \end{bmatrix} \begin{Bmatrix} u^1 \\ u^{12} \end{Bmatrix} = \begin{bmatrix} B_1 & B_{12} \end{bmatrix} \begin{Bmatrix} t^1 \\ t^{12} \end{Bmatrix}$$

$$\begin{bmatrix} A_2 & A_{21} \end{bmatrix} \begin{Bmatrix} u^2 \\ u^{21} \end{Bmatrix} = \begin{bmatrix} B_2 & B_{21} \end{bmatrix} \begin{Bmatrix} t^2 \\ t^{21} \end{Bmatrix} \tag{19}$$

for domain I and II respectively. Using continuity and equilibrium along the boundary given by

$$u^{12} = u^{21}$$

$$t^{12} = -t^{21} \tag{20}$$

(i.e. the displacements are continuous, while the tractions are equal and opposite on the common interface), Equation (19) can be combined as

$$\begin{bmatrix} A_1 & 0 & A_{12} & -B_{12} \\ 0 & A_2 & A_{21} & B_{21} \end{bmatrix} \begin{Bmatrix} u_1 \\ u_2 \\ u_{12} \\ t_{12} \end{Bmatrix} = \begin{bmatrix} B_1 & 0 \\ 0 & B_2 \end{bmatrix} \begin{Bmatrix} t_1 \\ t_2 \end{Bmatrix} \tag{21}$$

Combining Equation (21) with the boundary conditions allows all of the tractions and displacements along the boundaries and interface to be found. The scheme can be extended to more than two domains using the same technique without difficulty.

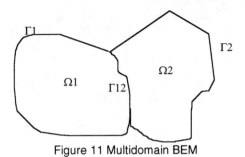
Figure 11 Multidomain BEM

The multidomain technique is useful for fracture problems in which the plane or cylinder containing the crack is placed on the interface between two domains. This allows opposite faces of the crack to be given identical positions in space. Giving the two faces of the crack identical positions in space using a single domain leads to a singular matrix.

Fracture Mechanics

The strain energy release rate is defined as the energy released per unit area for crack growth over a small area ΔA. The formula is

$$G_I = \lim_{\Delta A \to 0} \frac{1}{2\Delta A} \int_{\Delta A} (u_n^+ - u_n^-)\sigma_n dA$$

$$G_{II} = \lim_{\Delta A \to 0} \frac{1}{2\Delta A} \int_{\Delta A} (u_t^+ - u_t^-)\sigma_t dA$$

(22)

where GI and GII are the Mode I and Mode II strain energy release rates, respectively, ΔA is the crack extension area, (un+- un-) and (ut+- ut-) are the total crack opening and sliding, and σn and σt are the normal and shear stresses ahead of the crack. Strictly speaking, the stresses are those existing ahead of the crack tip before crack extension and the displacements are those behind the crack tip after crack extension so that the products in Equation (22) are taken for quantities at exactly the same position in space. Thus strict evaluation of Equation (22) requires BEM data for two different crack lengths. However, since the displacements just behind the crack tip are almost equal both before and after a small crack extension, Equation (22) can be approximated by using stresses and displacements for one crack length. This approximation is known as the modified crack closure integral, and it has been used successfully in finite elements (Rybicki and Kanninen, 1977, Sun and Jih, 1987) and three-dimensional boundary elements (Farris and Liu, 1993).

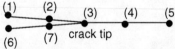
Figure 12 Nodes and elements near the crack tip used in modified crack closure integral

The modified crack closure integral is evaluated by substituting boundary element stresses and displacements into Equation (22). The stress and displacement shape functions are used in a closed

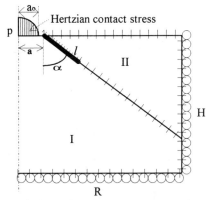

Figure 13 Boundary element mesh

form integration of Equation (22) leading to Mode I and Mode II strain energy release rates in terms of the nodal values of stress and displacement (Figure 12) as

$$G_I = \frac{u_n^{(1)} - u_n^{(6)}}{2}(\tfrac{2}{15}\sigma_n^{(3)} + \tfrac{1}{15}\sigma_n^{(4)} - \tfrac{1}{30}\sigma_n^{(5)}) + \frac{u_n^{(2)} - u_n^{(7)}}{2}(\tfrac{1}{15}\sigma_n^{(3)} + \tfrac{8}{15}\sigma_n^{(4)} + \tfrac{1}{15}\sigma_n^{(5)})$$

$$+ \frac{r^{(5)} - r^{(3)}}{r^{(5)} + r^{(3)}}[\frac{u_n^{(1)} - u_n^{(6)}}{2}(-\tfrac{1}{10}\sigma_n^{(3)} - \tfrac{1}{15}\sigma_n^{(4)}) + \frac{u_n^{(2)} - u_n^{(7)}}{2}(-\tfrac{1}{15}\sigma_n^{(3)} + \tfrac{1}{15}\sigma_n^{(5)})]$$ (23)

$$G_{II} = \frac{u_t^{(1)} - u_t^{(6)}}{2}(\tfrac{2}{15}\sigma_t^{(3)} + \tfrac{1}{15}\sigma_t^{(4)} - \tfrac{1}{30}\sigma_t^{(5)}) + \frac{u_t^{(2)} - u_t^{(7)}}{2}(\tfrac{1}{15}\sigma_t^{(3)} + \tfrac{8}{15}\sigma_t^{(4)} + \tfrac{1}{15}\sigma_t^{(5)})$$

$$+ \frac{r^{(5)} - r^{(3)}}{r^{(5)} + r^{(3)}}[\frac{u_t^{(1)} - u_t^{(6)}}{2}(-\tfrac{1}{10}\sigma_t^{(3)} - \tfrac{1}{15}\sigma_t^{(4)}) + \frac{u_t^{(2)} - u_t^{(7)}}{2}(-\tfrac{1}{15}\sigma_t^{(3)} + \tfrac{1}{15}\sigma_t^{(5)})]$$ (24)

where the superscripts represent the node point numbers. If the crack front is parallel to the z-axis, i.e. the crack is cylindrical in shape, equation (23)-(24) reduces to

$$G_I = \frac{u_n^{(1)} - u_n^{(6)}}{2}(\tfrac{2}{15}\sigma_n^{(3)} + \tfrac{1}{15}\sigma_n^{(4)} - \tfrac{1}{30}\sigma_n^{(5)}) + \frac{u_n^{(2)} - u_n^{(7)}}{2}(\tfrac{1}{15}\sigma_n^{(3)} + \tfrac{8}{15}\sigma_n^{(4)} + \tfrac{1}{15}\sigma_n^{(5)})$$

$$G_{II} = \frac{u_t^{(1)} - u_t^{(6)}}{2}(\tfrac{2}{15}\sigma_t^{(3)} + \tfrac{1}{15}\sigma_t^{(4)} - \tfrac{1}{30}\sigma_t^{(5)}) + \frac{u_t^{(2)} - u_t^{(7)}}{2}(\tfrac{1}{15}\sigma_t^{(3)} + \tfrac{8}{15}\sigma_t^{(4)} + \tfrac{1}{15}\sigma_t^{(5)})$$ (25)

Equations (23-25) assume that the crack extension length is the length of the element next to the crack tip on either side.

The relationship between stress intensity factor and strain energy release rate for axisymmetric problems is the same as the plane strain case. Thus, the stress intensity factors can be calculated from the strain energy release rates using

$$K_I = \left(\frac{G_I E}{1 - v^2}\right)^{1/2}$$

$$K_{II} = \left(\frac{G_{II} E}{1 - v^2}\right)^{1/2}$$ (26)

once G_I and G_{II} are calculated.

Cone Crack Analysis

The stress intensity factors around the cone crack are calculated and compared with the corresponding finite element results of Li and Hills (1991). Figure 13 shows the boundary element model for the stress analysis of the cone crack. The model used in the calculations has 60 quadratic elements for each domain. The element length near the crack tip is 0.1a. Figure 14 shows the calculated stress intensity factors obtained using the BEM and also the FEM results of Li and

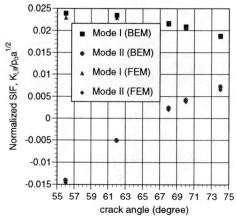

Figure 14 SIF's of cone cracks using BEM and FEM (Li and Hills, 1991), $l/a_0=1$, $a/a_0=1.05$

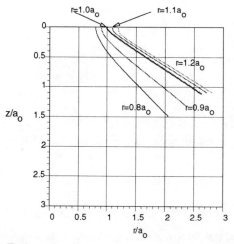

Figure 15 Trajectories for the minimum principal stress, $\nu=0.23$

Hills(1991). From this figure it is clear that our calculated values of the SIF's are within 3% of the values obtained by Li and Hills(1991). The close comparison validates the BEM results.

The cone crack angle used in Figure 14 is within the range of angles commonly observed in experiment. Many investigations have assumed that the crack grows perpendicular to the maximum principal stress induced by Hertz contact in the uncracked media. Trajectories of the minimum principal stress for $\nu=0.23$ are shown in Figure 15. The intermediate principal stress acts in the θ-direction so that the maximum principal stresses act perpendicular to the lines in Figure 15. These trajectories predict that the initial ring crack would form perpendicular to the surface before turning to grow at an angle into a cone crack. Since the length of growth perpendicular to the surface is usually small, the cone crack has been modeled as growing along a straight line at an angle to the surface. The largest tensile stress in the body is σ_r at r=a and z=0, and its value is

$$\sigma_r = \tfrac{1}{3}(1-2\nu)p_0 \tag{27}$$

Hence as the load is applied, ring cracks should occur just at the edge of contact. However, Johnson et al (1973) have shown from a stress analysis that includes the elasticity of the indenter that the location of this maximum value of σ_r may occur just outside of the contact; this is consistent with their observations of the location of ring crack occurrence as well as those of Zeng et al (1992). As the crack follows the stress trajectory during growth it will continue to have a positive Mode I stress intensity factor.

The BEM analysis has quantified one feature of the growth of the ring/cone crack system which has been discussed previously (Mouginot and Maugis, 1985). If a ring crack forms at the surface and does not propagate unstably, immediately, into a cone crack, then as the indenter load is increased, the ring crack will eventually be encompassed within the circle of contact of the surface. Under such conditions, the SIF calculation of Figure 16 shows that the growth of the ring crack into the sub-surface will be arrested. The negative Mode I stress intensity factors are physically not realistic but

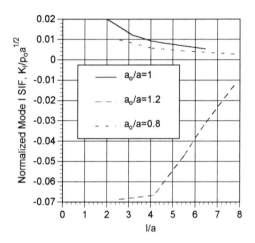

Figure 16 Normalized stress intensity factors
for Hertzian cone cracking, $\alpha=62^\circ$, and
$\nu=0.23$

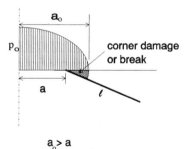

$a_o > a$

Figure 17 Corner breaking or damage
caused by indentation

indicate that for this configuration the crack would close and crack growth is arrested. Also it can be seen from Figure 16 that for a given contact size, the stress intensity factor decreases with increasing cone crack length when the ring crack size is larger than the contact size.

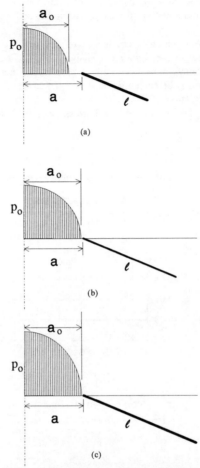

Figure 18 Continual cone cracking scheme:(a) contact patch size is smaller than cone crack radius at surface; (b) contact patch size equals cone crack radius at surface; (c) increasing load without increasing contact patch size

Experimental observations indicate that the cone crack continues to grow under increasing load even if the contact radius predicted by Hertzian analysis of uncracked solids is greater than the ring crack radius. This violates the stress intensity factor calculation in the previous paragraph. It is also found that continual loading leaves some permanent damage around the ring crack area (Figure 17). This damage changes the smooth Hertzian contact. Assuming the damage restricts the contact to occur inside the ring crack, a continual loading scheme is suggested for use as a modified boundary condition in the boundary element model. This scheme is illustrated in Figure 18. Three stages are defined in the continual loading. The three stages are (a) ring crack radius greater than contact size, (b) the ring crack radius equal to the contact size with the contact being Hertzian, and (c) non-Hertzian contact with the contact size equal to the ring crack radius. The relevant parameters obey the following relationships:

$$a^{(c)} = a^{(b)} = a^{(c)}$$
$$a_o^{(c)} < a_o^{(b)} = a_o^{(c)}$$
$$p_o^{(a)} < p_o^{(b)} < p_o^{(c)}$$
$$l_o^{(a)} < l_o^{(b)} < l_o^{(c)} \qquad (28)$$

where superscripts (a), (b), and (c) represent the loading stages respectively. In stage (c), the contact pressure increases to be larger than that given by the conventional Hertz contact equation. Assuming these stages leads the prediction of continual crack growth.

From the fracture toughness of the soda-lime glass and the stress intensity factors calculated using the boundary element method, one can obtain the corresponding normalized critical crack length. The contact length, a_0, is obtained using Hertz contact theory for stages (a) and (b)

$$a_o = \left(\frac{3PR}{4E^*} \right)^{1/3} \qquad (29)$$

where P is the indentation load, R is the indenter radius, E^* is the effective Young's modulus given by

$$\frac{1}{E^*} = \frac{1 - v_g}{E_g} + \frac{1 - v_s}{E_s} \qquad (30)$$

E and v are the Young's modulus and Poisson's ratio and the subscripts s and g refer to steel and glass, respectively.

If it is assumed that crack growth is arrested when the stress intensity factor is smaller than the fracture toughness, then the cone crack length can be predicted as a function of load. The predicted values using the modified continual loading scheme are shown in Figure 3 and Figure 4 for different ring crack radii. The predicted values are close to those measured experimentally for low indentation load. The radius of the surface trace of the ring/cone crack is difficult to measure experimentally due to damage of the glass surface, so that the predictions are made for a range of these radii. It is found that the predicted crack length decreases as the ring crack radius increases. This decrease is less than 5% for a 25% increase in ring crack radius. Thus different estimates of the ring crack radius does not significantly change the cone crack length prediction. The stress intensity factor calculation using a reasonable ring crack radius gives good agreement between predicted and measured cone crack length. The difference between the BEM predictions and experiment for higher loading might be due to extra damage initiated during contact and additional cracking initiated by indentation such as median cracks.

CONCLUSION

The relationship between the Hertzian cone crack length and the indentation load in soda-lime glass has been measured experimentally and predicted using axisymmetric boundary elements. Experimental observation shows that changing the ball radius by a factor of two has a slight effect on the cone crack length while the low values of the Mode II stress intensity factor explain the near straight profile of the cone crack. The estimated cone crack length-load relationship agrees well with experimental values for the load range in which the cone crack is fully developed and there is no major influence from additional crack systems. The initial ring crack radius assumption has a small effect on stress intensity factor calculation provided that it is assumed that the contact radius is not greater than the ring crack radius. Thus, the simple boundary element model and the modified continual loading scheme accurately predict the cone crack fracture growth for a fairly wide load range. The use of cone cracking as a fracture toughness test and as a model for a material removal mechanism in fine finishing can now be pursued with more confidence. However, future work should pursue BEM analysis of the actual contact problem for the cracked solids rather than assuming a Hertzian pressure distribution as is done here.

Acknowledgment - This research is supported in part by the National Science Foundation through grants MSS9057082, Dr. J. Larsen-Basse, Program Director and DDM 9057916, Dr. B. Kramer, Program Director.

REFERENCES

Abramowitz, M. and Stegun, I.A 1972, *Handbook of Mathematical Functions*, Dover, New York, 337

Bakr, A.A., 1983, "Boundary Integral Equation Analysis of Axisymmetric Stress and Potential Problems," PhD Thesis, Imperial College, University of London

Bakr, A.A., 1985, *The Boundary Integral Equation Method in Axisymmetric Stress Analysis Problems*, Springer-Verlag,

Brebbia, C. A. 1980, *The Boundary Element Method for Engineers*, 2nd Edn. Pentech Press, London; Halstead Press, New York

Brebbia, C.A., Telles, J.C.F., and Wrobel, L.C. 1984, *Boundary Element Techniques*, Springer-Verlag, Berlin

Chen, S.Y., and Farris, T.N. 1993, "Boundary Element Crack Closure Calculation of Axisymmetric Stress Intensity Factors", submitted to *Computers and Structures*

Cruse, T.A., Snow, D.W., and Wilson, R.B., 1977, "Numerical Solutions in Axisymmetric Elasticity," *Comp. and Struct.*, 7, pp445-451

Erdelyi, A. et al. 1953, *Higher Transcendental Functions*, Bateman Manuscript Project, 1, McGraw-Hill, New York

Farris, T.N. and Liu, M., 1993, "Boundary Element Crack Closure Calculation of Three-Dimensional Stress Intensity Factors," *International Journal of Fracture*, 60, pp 33-47

Frank, F.C., and Lawn B. R. 1967, "On the Theory of Hertzian Fracture", 299, A, pp. 291-306

Johnson, K.L., O'Connor, J.J., and Woodward, A.C. 1973, "The Effect of the Indenter Elasticity on the Hertzian Fracture of Brittle Materials", *Proc. R. Soc. Lond.*, A. 334, 95-117

Kermanidis, T., 1975, "A Numerical Solution for Axially Symmetrical Elasticity Problems," *Int. J. Solids Struct.*, 11, pp493-500

Li, Y., and Hills, D.A. 1991, "The Hertzian Cone Crack," *Transactions of the ASME, Journal of Applied Mechanics*, 58, pp120-126

Love, A.E.H. 1944, *A Treatise on the Mathematical Theory of Elasticity*, Dover, New York

Mayr, M., 1976, "The Numerical Solution of Axisymmetric Elasticity Problems Using an Integral Equation Approach," *Mech. res. Com.*, 3, pp393-398

Mouginot, R. and Maugis, D., 1985, "Fracture Indentation Beneath Flat and Spherical Punches," *Journal of Materials Science*, Vol 20, pp 4354-4376

Rybicki, E.F. and Kanninen, M.F., 1977, "A Fnite Element Calculation of Stress Intensity Factors by a Modified Crack Closure Integral," *Engineering Fracture Mechanics*, Vol 9, pp 931-938

Sun, C.T., and Jih, C.J., 1987, "On Strain Energy Release Rates for Interfacial Cracks in Bi-Material Media," *Engineering Fracture Mechanics*, 28, pp 13-20

Warren, R., 1978, "Measurement of the Fracture Properties of Brittle Solids by Hertzian Indenters," *Acta Metallurgica*, 46, pp 1759-1769

Zeng, K., Breder, K., and Rowcliffe, D.J. 1992a, "The Hertzian Stress Field and Formation of Cone Cracks-I. Theoretical Approach", *Acta metall. mater.* 40, No. 10, 2596-2600

Zeng, K., Breder, K., and Rowcliffe, D.J. 1992b, "The Hertzian Stress Field and Formation of Cone Cracks-II. Determination of Fracture Toughness", *Acta metall. mater.* 40, No. 10, 2600-260

AD-Vol. 36, Fatigue and Fracture of Aerospace Structural Materials
ASME 1993

THROUGH-THE-THICKNESS STRENGTH MEASUREMENT
OF COMPOSITE LAMINATES USING AN
L-SHAPED CURVED BEAM SPECIMEN

Harold G. Allen, Kunigal N. Shivakumar,
and Vishnu S. Avva
Department of Mechanical Engineering
North Carolina A&T State University
Greensboro, North Carolina

ABSTRACT

Through-the-thickness strength of laminated composites is an important parameter in design of structures subjected to out-of-plane loads, structural joints, and attachments. To measure this strength a 2x2 inch (51x51 mm) L-shaped curved beam specimen and a tension loading fixture are developed. The use of standard tension test machine and introduction of load at the specimen mid-thickness were the advantages of this specimen. Test and analysis of 16, 24, and 32 ply thick curved laminates were conducted. Measured through-the-thickness strengths of 16, 24, and 32 ply laminates were 82, 67, and 35 MPa, respectively. Results of 16 and 24 ply specimens agreed very well with the data in literature for a longer size specimen and different loading fixture. A lower bound through-the-thickness strength value of 82 MPa for laminate thickness less than 2 mm and 67 MPa for thickness between 2 and 3 mm maybe used in structural designs. For thick laminates, the through-the-thickness strength value measured from the 90 deg tension specimen (58 MPa) maybe used. Decrease in measured strength with an increase in laminate thickness is due to larger volume of material in a high stress state and the probability of large number and size of defects in that volume of material.

INTRODUCTION

Test methods for measuring in-plane strengths -- longitudinal, transverse, and shear strengths -- are well established and the procedures are well documented in American Standard method for Testing and Materials handbook. But no standard has been developed to measure the through-the-thickness strength of laminated composite materials. Through-the-thickness strength is an important parameter in design of structures subjected to out-of-plane loads, structural joints, attachments, etc. . Several attempts have been made in the literature to measure the transverse strength (some times it is referred to as the interlaminar strength) of laminates and to establish a test specimen and the procedure. The two types of specimens used were 90 deg laminate tension specimen [1-3] and L-shaped curved beam specimen [4-7]. The transverse tension specimen had a number of drawbacks, including the quality of fabrication, applying uniform stress, and stress concentration due to specimen taper. Furthermore, transverse strength of unidirectional laminated composite may not be same as the through-the-thickness strength of laminated composites used in structural applications, because of differences in the triaxial stress constraint. Perhaps, the transverse tension specimen data may provide a lower bound strength value.

After Ko and Jackson [4] showed that thick curved multilayer composite specimens develop through-the-thickness (radial) stresses that are high enough to cause the transverse failure of laminates, L-shaped curved beams specimens were proposed [4-7]. This specimen with different types of loading arrangement was used to measure the transverse strength (delamination on set) and delamination growth of laminated composites. Although test specimens were successful to a varied degree, the problem of load introduction and specimen size were not addressed satisfactorily. The specimen size of textile composites are restricted by the braiding mandrel or the weaving machine. To use the same size specimen for laminated as well as textile composites, the specimen size in this study was restricted to about 2 inches (51 mm).

Objectives of this paper are to develop a 2x2 inch L-shaped curved beam specimen and appropriate load fixture so that it can be tested in a standard tension test machine, measure the through-the-thickness tension strength of unidirectional laminated composite, and compare the measured strength data with those reported in the literature. The proposed specimen is an L-shaped curved beam with arm lengths of 2 inches (51 mm) and inner radius at the curved section is 0.25 inches (6.4 mm). This specimen is similar to that which was proposed in reference 7, but shorter in length. As mentioned previously, the arm length was selected based on the fabrication restriction of textile composite materials. This same specimen is proposed to be used to test the through-the-thickness tension strength of braided and woven composites.

L-shaped curved beam specimen configuration and fabrication procedure, and the load fixture are described. Three laminate thickness, namely, 16, 24, and 32 ply are fabricated and tested. Finite-element analysis and beam theory solution are used to calculate the transverse tension strength and then results are compared with those reported in the literature for different specimen configurations.

SPECIMEN CONFIGURATION AND LOADING FIXTURE

Test Specimen

Proposed L-shaped curved beam specimen is shown schematically in the figure 1. The specimen length is about 2 inches (51 mm), this length was arrived based on the size restriction of textile composites. The inner radius (r_i) of the beam at the curved section is 0.25 inches (6.4 mm). The specimen width (w) is 1.5 inches (38 mm). Three thicknesses (t) were chosen, namely, 16, 24, and 32 plies. The material is AS4/3501-6 graphite/epoxy composite with lamina properties:

E_{ll} = 19 Msi (131 GPa) E_{tt} = 1.89 Msi (13 GPa), G_{lt} = 0.93 MSI (6.4 GPa)
v_{lt} = 0.34,   and v_{tt} = 0.35

Where subscripts l and t refer to longitudinal and transverse directions of the fibers, respectively. Unidirectional laminated specimen was selected to prevent matrix cracks and edge delaminations.

The curved laminated specimen was fabricated by laying up 12 inches (305 mm) strips of unidirectional plies over the corner of a solid aluminum mold. The radius of the corner was 0.25 inches (6.4 mm). Each of the straight arm was about 2.5 inches (64 mm) long. The stacked plies and the mold were vacuum bagged and cured in autoclave according to the material manufacturers instructions. The specimen were then machined to 1.5 inches (38 mm) wide and 2 inches (51 mm) long from the 12x2.5 inches (305x64 mm) laminate. Two edges of the specimen were polished and the exact width and the thickness were measured using a caliper. Specimen edges were painted white with a water based typewriter correction fluid to identify the location of the transverse tension failure. Five specimens were manufactured for each of the three thicknesses.

Loading Fixture

A special load fixture was fabricated so that the specimen can be tested in a standard tension test machine and the load is introduced at the mid-thickness of the specimen. Figure 2 shows schematically the specimen attachment fixture. Ends of the specimen are clamped using two reusable aluminum tabs and two 0.25 inches (6.4 mm) high strength bolts. The two bolts were spaced 0.75 inches (19 mm) apart in the width direction and were tightened by about 50 inch-pound torque. Thickness of the top tab at loading end was equal to the sum of thicknesses of the two tabs at the bolted end and 24 ply laminate (3.2 mm). The specimen was loaded through a 0.5 inches (13 mm) diameter pin, which was attached to the test machine through a U-frame universal joint (see Fig. 3). The pin axis nearly coincided with the mid-plane of the 24 ply thick laminate (see Fig. 2). Note that the pin axis was shifted by about ±4 ply (0.5 mm) for 16 and 32 ply laminates. Although this shift could have been adjusted using shims, it was not done in the present test. The effect of shift in the load axis was assumed to be small and was neglected.

TEST PROCEDURE AND RESULTS

The specimen and the loading fixture assembly (see Fig. 3) was mounted on a hydraulically controlled tension test machine. A small preload (about 5 lbs) was applied on the specimen to measure the distance between the load axis and outer edge of the specimen, s (See Fig. 2). This procedure was repeated for all specimens. The measured value of s was used in finite element modeling of the specimen loading arm. Tests were conducted under stroke displacement controlled at 1 mm/min. The load and stroke displacements were digitally recorded. The transverse tension failure was recorded as a sudden decrease in the load and also as a loud "pop" noise. The load drop was more than 20% of the peak load for most of the specimen. The load at the incipient of delamination represented the transverse tension failure load and it was used in calculating the tension strength. Observation on edges of the specimen showed a delamination near the mid thickness of the specimen, where the transverse tension stress was highest. Subsequent loading on the specimen showed multiple steps of load decreases associated with the development of multiple delaminations on either side of the initial delamination. All fifteen specimens were tested and the load displacements were recorded.

Figures 4 through 6 show the load-displacement curves for 16, 24, and 32 ply laminates. To improve the clarity of figures, only four specimen data are shown for each of the thickness. Initiation of through-the-thickness tension failure are shown by a sudden drop in the load. The load at the incipient of the transverse tension failure is the largest value and the specimen did not bear a load larger than this value until the specimen became almost straight. The failure load was used in the analysis to calculate the through-the-thickness strength.

The load-displacement curves for 32 ply laminate showed nonlinear response due to initiation of premature failures at resin rich pockets. Quality of 32 ply laminates were poor compare to 16 and 24 ply laminates. Molding of 32 ply laminate took more time and it was difficult. While molding the laminate, each ply at a time, ply crimps and waviness were almost unavoidable after about 25 plies were molded. These crimps and waviness contributed to development of resin rich pockets. No such problems were encountered in fabricating 16 and 24 ply laminates.

ANALYSIS

Two-dimensional (2-D) finite-element (F-E) analysis of the L-shaped curved beam specimen was conducted using an in-house finite element code FEM2D8. Detailed stress field at the curved region of the specimen was calculated. The maximum radial stress at the curved section was compared with the beam solution [4].

Finite-Element Analysis

An in-house 2-D finite-element code FEM2D8, based on 8-node isoparametric element, was used to conduct the stress analysis of the L-shaped curved beam specimen and the loading fixture. The loading fixture was idealized as an equivalent arm length from the measured value of s, during the test. Because the specimen and the loading were symmetric, only top-half of the specimen was modeled. The model was idealization had 15 elements through the thickness and one element at every one degree in the circumferential direction. A transition was made to a coarse mesh to model the straight loading arm (see Fig. 7). The model had 945 elements, 2,992 nodes, and 5,984 degrees-of-freedom. The load P was applied at the right end and mid-thickness of the model. A small offset in the load axis for 16 and 32 plies was neglected. Symmetric boundary condition on y = 0 plane and zero x-displacement at the load point were imposed. The plane-stress analysis was conducted by setting the load, P, to be unity. Calculated stresses were scaled by the failure load measured in the test. The analysis was repeated for all specimens using the measured values of the laminate thickness and s.

Because the deflection at the incipient of the transverse tension failure was small, the effect of large deflections was neglected.

Beam Theory Solution

Kedward et al. [4] proposed an elementary beam theory method to determine the maximum radial stress in a circular curved composite sections. The maximum stress in the circular curved section of the proposed L-shaped beam with an end loading is given by the equation

$$(\sigma_r)_{max} = \frac{3PL\,Cos(\omega)}{2wt\sqrt{r_i\,(r_i + t)}}$$

Where ω is the angle between the laminate arm and the x-axis (see Fig. 1). This equation accounts for the maximum stress not being at the mid-thickness of the specimen.

Analytical Results

Figure 8 show distribution of normalized radial tension stress along the x-axis for three typical specimen thicknesses. The abscissa represent the normalized radial distance measured from the inner surface of the specimen. The ordinate represent the radial stress σ_r normalized by the applied load P and specimen width and thickness. Value of $(r-r_i)/t$ variation between zero and one representing the inner to outer surfaces of the specimen. Results for 16, 24, and 32 ply laminates calculated from the F-E analysis and the beam theory solution are presented. Since the beam solution gave only the maximum stress, it is shown by a horizontal line. Maximum radial stress occurred, when measured from the inner surface, at 47% thickness for 16 ply laminate and at 40% thickness for 24 and 32 ply laminates. The test data also showed that the tension failure occurred in this region. The beam theory solution [4] is about 12.6%, 6.9%, and 1.1% smaller than finite element solution for 16, 24, and 32 ply laminates, respectively. The beam theory solution appears to be accurate for thick laminates (32 plies and more).

Variation of tangential stress σ_θ along the x-axis is shown in the figure 9. Tangential stress varies nonlinearly through the thickness and has the maximum value at the inner surface. The ratio of maximum radial to tangential stress is 0.06, 0.08, and 0.11 for 16, 24, and 32 ply laminates, respectively. A decrease in ratio of radial to tangential stresses with decrease in the laminate thickness indicate that a likelihood of transverse tension failure of specimens thinner than 16 plies is low. Therefore, the thin laminates may not fail by transverse tension. In contrast, the likelihood of failure of thick curved laminates by transverse tension is high.

THROUGH-THE-THICKNESS TENSION STRENGTH
Through-the-thickness strength of all specimens were calculated from the measured failure load and the finite-element analysis. Calculated strength valued for 16, 24, and 32 plies are shown in figure 10. Four specimen data for each thickness are presented. The solid symbol represent the mean value, it is 82, 67, and 35 MPa for 16, 24, and 32 ply laminates, respectively. The laminate strength is proportional to magnitude of the stress, the volume of material under high stress, and the number and size of defects in that volume of material. Among the 16 and 24 ply laminates, volume of material under high stress and the associated number and size of defects contributed to the difference in the measured strength values. In case of 32 ply laminates, in addition to large volume of material under high stress and random defects, quality of the fabrication also contributed to the lower value of the strength. Because the thickness of secondary structures used in aircraft's and space is less than 0.1 inches (2.5 mm) and the fabrication problems experienced for 32 ply laminates may not be encountered. In such applications, specimen thicknesses of specific structural components may be fabricated and the corresponding strengths are measured. In absence of fabrication and testing facilities, a conservative value of 82 MPa may be used for thicknesses less that 2 mm and 67 MPa for thicknesses between 2 and 3 mm.

Figure 11 compares the through-the-thickness strength values from the present study with data from references 3 and 7. The data in reference 3 was for 90 deg tension test specimen and the reference 7 was for 25 mm wide and 80 mm long L-shaped curved beam specimen. The loading arrangement in reference 7 was different from that was used in the present study. Although the specimen widths and loading arrangements were different, the present results for 16 and 24 ply laminates agreed very well with data in reference 7. The strength values of 16 and 24 ply laminates are higher than that of the 90 deg tension data (58 MPa), because 90 deg specimen has no triaxial stress constraint and the whole cross-section is subjected to uniform high stress state. However, all curved beam laminates considered, develop a through-width constraint similar to plane strain constraint (due to a large ratio of width to thickness of the specimen), and only the mid-thickness of the specimen is highly stressed. Therefore, for thick laminates, 90 deg tension specimen strength data would be a good representative value.

CONCLUDING REMARKS
A 2x2 inch (51x51 mm) L-shaped curved beam specimen and a tension loading fixture are proposed to measure the through-the-thickness strength of laminated and textile composites. The use of standard tension test machine and introduction of load at the specimen mid-thickness were the advantages of the proposed specimen. Test and analysis (finite element and curved beam theory) of 16, 24, and 32 ply thick curved laminates were conducted. Quality of 16 and 24 ply laminates was very good but the 32 ply laminates was poor due to fabrication difficulties. Measured through-the-thickness strengths of 16, 24, and 32 ply laminates were 82, 67, and 35 MPa, respectively. These values were based on the average of four specimens data. Results of 16 and 24 ply specimens agreed very well with data in the literature for a longer size specimen and different loading fixture. In the absence of any other strength data, a lower bound through-

the-thickness strength value of 82 MPa for thickness less than 2 mm and 67 MPa for thickness between 2 and 3 mm maybe used in structural designs. For thick laminates, the through-the-thickness strength value measured from the 90 deg tension specimen (58 MPa) maybe used. Decrease in measured strength with an increase in laminate thickness is due to larger volume of material in a high stress state and the probability of large number and size of defects in that volume of material.

Acknowledgments
The authors acknowledge the financial support of National Aeronautics and Space Administration, Washington, DC by a Grant No. NAGW-1331 to the Mars Mission Research Center, a joint program between NC State University and NC A&T State University. The computational support was provided by the North Carolina Supercomputing Center, Research Triangle Park, NC.

REFERENCES
1. Martin, R. H. and Sage, G. N. : "Prediction of the Fatigue Strength of Bonded Joints Between Multi-Directional Laminates of CFRP," Composite Structures, Vol. 6, 1986, pp. 141-163.
2. Lagace, P. A. and Weems, D. B., "A through-the Thickness Strength Specimen for Composites," Test Methods for Design Allowables for Fibrous Composites: 2nd Vol., ASTM STP 1003, C. C. Chamis, Ed., American Society for Testing and Materials, Philadelphia, 1989, pp. 197-207.
3. O'Brien, T. K. and Salpekar, S. A. : "Scale Effects on the Transverse Tensile Strength of Graphite Epoxy Composites," 11th ASTM Symposium on Composite Materials: Testing and Design, Pittsburgh, Pennsylvania, May 5-6, 1992.
4. Ko, W. L. and Jackson, R. H. :"Multilayer Theory for Delamination Analysis of Composite Curved Bar Subjected to End Forces and End Moments," In Composite Structures 5; Proc. of the Fifth International Conference, Paisley, Scotland, July 24-26, 1989.
5. Hiel, C. C., Sumich, M., and Chappell, D. P. :"A Curved Beam Test Specimen for Determining the Interlaminar Strength of a Laminated Composite," Journal of Composite Materials, Vol. 25, July 1991, pp. 854-868.
6. Kedward, K. T., Wilson, R. S., and McLean, S. K. :"Flexure of Simply Curved Composite Shapes," Composites, Vol. 20, No. 6, Nov. 1989, pp. 527-536.
7. Jackson, W. C. and Matrin, R. H. : "An Interlaminar Tension Strength Specimen," 11th ASTM Symposium on Composite Materials: Testing and Design, Pittsburgh, Pennsylvania, May 5-6, 1992.

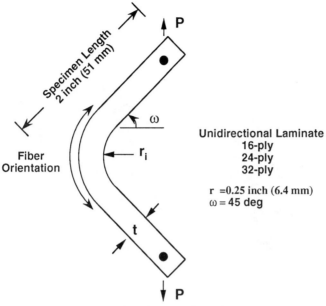

Specimen Length
2 inch (51 mm)

P

ω

r_i

Fiber
Orientation

t

P

Unidirectional Laminate
16-ply
24-ply
32-ply

r = 0.25 inch (6.4 mm)
ω = 45 deg

Figure 1.- Specimen configuration

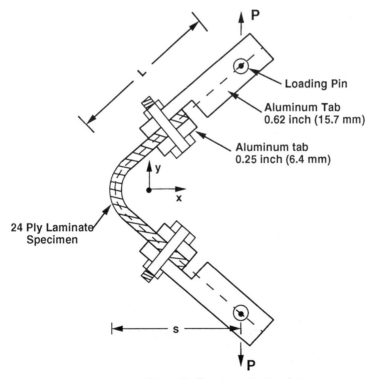

P

L

Loading Pin

Aluminum Tab
0.62 inch (15.7 mm)

Aluminum tab
0.25 inch (6.4 mm)

y

x

24 Ply Laminate
Specimen

s

P

Figure 2.- Specimen loading fixture

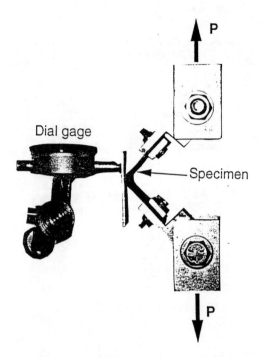

Figure 3.- Test specimen and loading fixture assembly.

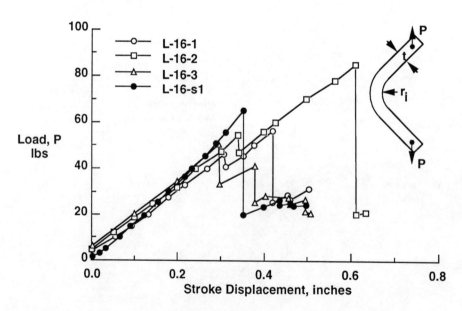

Figure 4.- Load-displacement plot of 16 ply thick specimens.

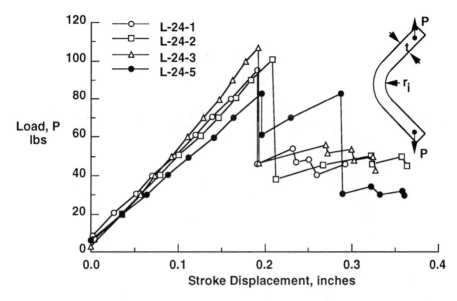

Figure 5.- Load-displacement plot of 24 ply thick specimens.

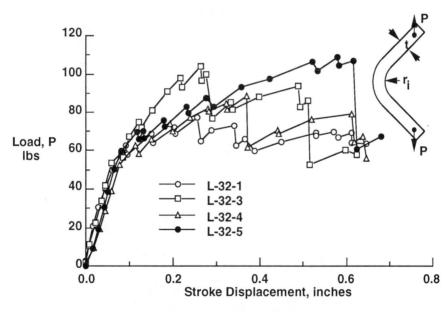

Figure 6.- Load-displacement plot of 32 ply specimens.

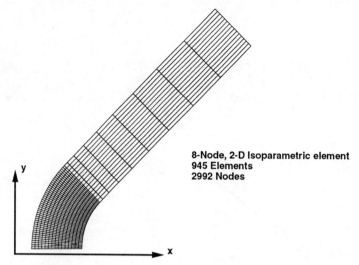

8-Node, 2-D Isoparametric element
945 Elements
2992 Nodes

Figure 7.- Finite-element idealization of the specimen and the equivalent loading arm.

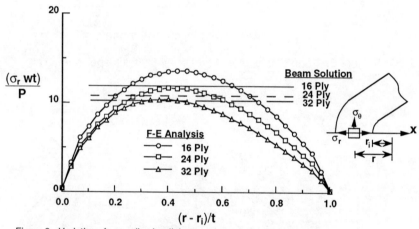

Figure 8.- Variation of normalized radial stress through-the-thickness of the specimen.

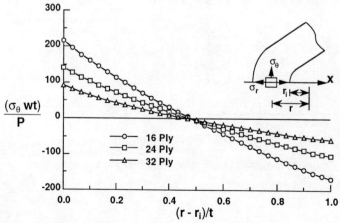

Figure 9.- Variation of normalized tangential stress through-the-thickness of the specimen.

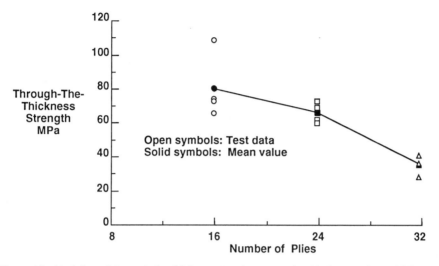

Figure 10.- Variation of through-the-thickness tension strength with the specimen thickness.

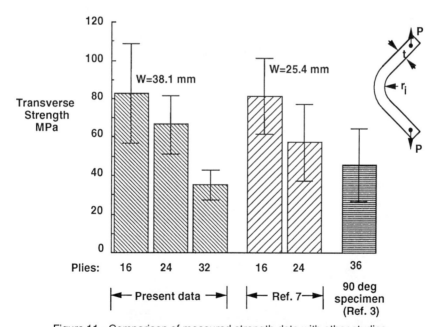

Figure 11.- Comparison of measured strength data with other studies.

AD-Vol. 36, Fatigue and Fracture of Aerospace Structural Materials
ASME 1993

THE IMPACT OF CONSTITUTIVE
MODELING ON LIFE PREDICTION

Donald C. Stouffer
Department of Aerospace Engineering
and Engineering Mechanics
University of Cincinnati
Cincinnati, Ohio

Gary Salemme
General Electric Aircraft Engines
Evendale, Ohio

ABSTRACT

This paper shows the relationship between constitutive modeling and life prediction technology. In particular, it is necessary to know the correct stress and strain components for the realistic prediction of crack initiation and crack growth. This paper focuses on recent results for the hysteresis loops in nonproportional tension-torsion loading and uniaxial thermomechanical loading. In both cases extra hardening is present that is not observed in proportional or isothermal loading. Constitutive models are proposed for the extra hardening from explanations of the microstructural events. Thus, this paper contains a brief description of the key deformation mechanisms present in metals, and system of state variable equations that can model most of the usual time dependent effects present at high temperatures or high strain rates. Then two modifications are introduced for multiaxial and thermomechanical fatigue, and there are several comparisons between experimental and model results.

1 INTRODUCTION

The fatigue life of a component is generally accepted to be made up of a crack initiation phase and a crack growth phase. These calculations are not an easy process to implement in real structures in the presence of real operating conditions. In most structures the stress and strain distributions are not uniform, and in regions of sharp changes in geometry, such as the fiber-matrix interface in metal matrix composites, there are relatively high local stress. If the stresses or temperature are high enough, an inelastic deformation history results making the prediction of crack initiation or crack growth more challenging.

The crack initiation life is frequently estimated in metals by the strain life approach. As the name implies, the approach depends on the correct local elastic-plastic strain histories for life prediction. For example, the Morrow (1968), Manson Halford (1981), and Smith Watson Topper (1970) equations all require the stress range, strain range and mean stress to determine the life in uniaxial loading. However time dependent inelastic effects such as strain rate dependence, stress relaxation, strain recovery and creep can make it difficult to obtain the correct input data for these models. Bannantine et. al. (1990) recently reported deficiencies in all three of the models even for strain rate independent elastic-plastic deformation histories.

The presence of elastic plastic behavior is also very important in crack growth. A plastic zone develops around the crack tip and in the wake of the crack in elastic plastic fracture. This plastic zone changes the local stresses and strains in the vicinity of the crack and raises doubts about the validity of using stress intensity factors determined for linear elastic fracture. The Wheeler

(1972) and Willenborg et. al. (1971) models for the overload retardation effect in variable amplitude loading require an estimate of the plastic zone size and stress range inside the plastic zone. Then there is also a philosophical issue about the size of the crack at the transition between the initiation and growth phases.

The condition of life prediction in multiaxial loading is even more challenging Both crack initiation and crack growth in multiaxial loading require a thorough understanding of material behavior. For example, there is a big difference in the response to proportional and nonproportional tension-torsion loading. In strain controlled proportional loading the axial strain, $\varepsilon(t)$, and shear strain, $\gamma(t)$, at any time are defined by

$$\varepsilon(t) = f(t)\,\varepsilon_o\,, \quad \text{and} \quad \gamma(t) = f(t)\,\gamma_o, \tag{1}$$

where ε_0 and γ_0 are constants and f(t) is the same driving function of time in both equations. Thus the ratio of axial strain to shear strain is constant throughout the deformation history. In nonproportional loading the values of $\varepsilon(t)$ and $\gamma(t)$ are totally independent. The degree of nonproportionality can easily be controlled experimentally by defining

$$\varepsilon(t) = \varepsilon_o \sin \varpi t, \quad \text{and} \quad \gamma(t) = \gamma_o \sin(\varpi t + \beta) \tag{2}$$

where ϖ is a cyclic frequency and β is a phase shift. Once again ε_0 and γ_0 are reference values for axial and shear strain, respectively. If $\beta = 0$, then the axial and shear strain are proportional, and if $\beta = \pi/2$ the axial and shear strains are 90 degrees out of phase and the test is nonproportional.

An example of the effect of nonproportional loading was demonstrated in copper at room temperature by D, McDowell (1983). The response to proportional tension-torsion loading followed two nonproportional load histories is shown in Figure 1. The nonproportional strain history given in Block 2 of Figure 1 was constructed to show the effect of a simple change in loading history, and Block 3 shows the effect of a highly nonproportional history. The results show that the amount of hardening increases as the degree of nonproportionality increases. The result also indicates that nonproportional response can not be predicted from uniaxial testing alone, thus many of the three dimensional constitutive models developed as an extension of uniaxial testing may not be accurate.

A similar effect can also be observed on thermomechanical cycling. It will be shown in Section 5 that the response to simultaneous thermal and mechanical cycling can not be extrapolated for isothermal testing when the thermal history involves deformation by two different deformation

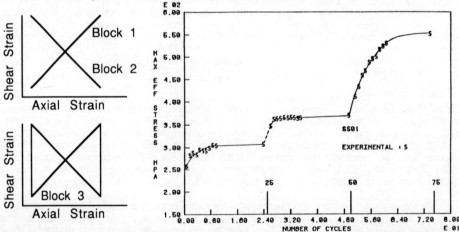

Figure 1. Sequence of load histories (left) and effective (Von Mises) stress strain curve as function of the load history for AISI Type 304 Stainless Steel at room temperature. (D. Mc Dowell, 1983).

mechanisms. In this case the microstructures for the two deformation mechanisms interact to produce effects that can not be observed by isothermal testing.

The objective of this paper is to discuss the extra hardening effects that may be present in metals in nonproportional loading and thermomechanical loading, and to present some ideas for constitutive models that account for these effects. The development starts with a brief discussion of the deformation mechanisms that occur in metals to provide a physical basis for the modeling. A state variable model is then presented. The results are next extended to multiaxial and thermomechanical loading where mechanism interaction effects are observed.

2 DEFORMATION MECHANISMS IN METALS

Inelastic deformation occurs in metals primarily due to slip and climb. When the temperature is less than approximately $0.5T_M$, where T_M is the absolute melting temperature, crystalline metals deform primarily due to the propagation of dislocations through the lattice (slip). When a shear stress is applied to a metal crystal, slip occurs if the shear stress exceeds a critical value. The atoms move an integer number of atomic distances along the slip plane and the crystal is perfectly restored if the motion is uniform. Since the crystal remains a crystal after slip, slip in metal crystals occurs on special planes of high atomic density in close pack directions. Under these conditions the shear stress propagating the dislocation corresponds to the minimum stress that can produce an identical lattice structure in the slipped position. In polycrystalline metals with a large number of grains at arbitrary orientation, the slip occurs on the slip planes and slip directions oriented closest to the maximum shear stress. For example, in a tensile test slip occurs on the slip planes oriented closest to 45 degrees from the tensile axis.

At higher temperatures deformation by dislocation climb becomes more important. Dislocation climb occurs by the diffusion of atoms and vacancies to or away from the site of an edge dislocation (an extra plane of atoms that terminates inside the crystal lattice). This results in the dislocation moving in a direction normal to the slip plane. The addition or subtraction of atoms to or from an edge dislocation will result in jogs in the dislocation line since diffusion of atoms is local and not expected to be uniform. Thus the dislocation line may lie in several slip planes simultaneously. This will result in an increase in the shear stress required to produce slip. Thus at high temperature two mechanisms can contribute to the observed inelastic strain, slip and climb; and, and as the temperature increases the potential for climb increases.

A major factor that controls the movement of dislocations is the interaction or intersection of dislocations with each other. As the network of dislocations increases, the complexity of the microstructure increases. The subsequent propagation of a dislocation through a network of dislocations requires a higher shear stress to maintain motion. An increase in the resistance to slip corresponds to strain hardening.

One of the most fundamental concepts to explain strain hardening is the interaction between individual dislocations in the same plane when they pile up at a barrier in the crystal lattice. Consider the propagation of dislocations on a slip plane under the action of a shear stress. The dislocation will be stopped at a barrier. A second dislocation following the same path will encounter the pinned dislocation and be repelled by forces on the atomic level. The dislocation pile ups produce a "back stress" which opposes the applied shear stress on the slip plane. At high temperature, as shown in Figure 2, slip in the reverse direction can result if the applied stress is reduced or removed. Thus, the net force producing slip is the difference between the shear stress and back stress. They are oriented on the slip plane in the slip direction and are tensor quantities.

Back stress results from the interaction between dislocations in the same plane. But since several slip planes may be active in a crystal at one time, dislocations propagating on different planes may intersect at various angles. Forcing a dislocation through another dislocation requires an element of work, and can produce discontinuous dislocations which require more energy to propagate. Thus an increase in drag (drag stress) also results from dislocation intersections. Drag stress is generally taken as a scalar quantity that embodies all scalar hardening effects.

Recovery is the opposite of hardening. It is associated with a reduction in the number of dislocations, a reduction in the stress required to propagate dislocations, and a reduction in the local

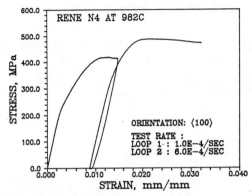

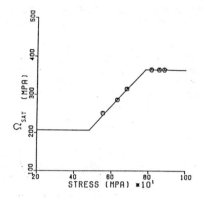

Figure 2. Strain rate effect and strain recovery in two minute hold period at zero stress in Rene N4 single crystal superalloy at 982 °C

Figure 3. Variation in the steady state value of the back stress with the creep stress for Rene` 80 at 760 °C.

energy level. There are two general types of recovery, static recovery and dynamic recovery. Static recovery occurs when there are no applied loads, and the movement of dislocations occurs from the interaction stresses between the dislocations themselves. Static recovery is more important at high temperatures when diffusion is present and dislocation mobility is assisted by thermal energy. Dynamic recovery occurs simultaneously with deformation. Dynamic recovery results from the formation of two substructures: subgrains at high temperature and cells at low temperature. Dynamic recovery acts to lower the dislocation density, distortion, and strain energy of the lattice structure. Thus the effective rate of strain hardening is reduced by dynamic recovery. Strain hardening and dynamic recovery are competing mechanisms during deformation.

3 STATE VARIABLE CONSTITUTIVE MODELING

The constitutive model of Ramaswamy et. al. (1984, 1986, 1990) and Stouffer et. al.(1990) is based on separating the total strain into the sum of elastic and inelastic components. This summation is solved for both loading and unloading, and the use of a separate unloading rule is not necessary such as in yield surface plasticity. The formulation is based on the use of three state variables, back stress, drag stress, and static thermal recovery. The back stress is motivated by the interaction of mobile dislocations with barriers such as grain boundaries, precipitates, and other dislocations. The back stress, Ω, is used to model the strain hardening observed in tensile response and the Bauschinger effect found in cyclic plasticity. Drag stress, Z, is included to model the effects of dislocations with particles or precipitates in the microstructure that generally result in cyclic hardening or softening. The static thermal recovery variable is included to account for the temperature driven effects present during creep. .

To begin, the total strain, ε_{pq}, is written as the sum of the elastic, ε_{pq}^{el}, inelastic, ε_{pq}^{inel}, and thermal strains, ε_{pq}^{th}, that is

$$\varepsilon_{pq} = \varepsilon_{pq}^{el} + \varepsilon_{pq}^{inel} + \varepsilon_{pq}^{th}. \tag{3}$$

The inelastic strain is the integral of the inelastic strain rate and assumed to be equal to the deviatoric inelastic strain since there is no inelastic volume change. The elastic and thermal strains are given by

$$\varepsilon_{pq}^{el} = \frac{1+\nu}{E} \sigma_{pq} - \frac{\nu}{E} \sigma_{kk} \delta_{pq} \quad \text{and} \quad \varepsilon_{pq}^{th} = \alpha \, \Delta T \, \delta_{pq} \tag{4}$$

where δ_{ij} is Kronecker Delta. The elastic modulus, E, Poisson Ratio, ν, and coefficient of expansion, α, are all functions of the current temperature, T, and ΔT is the deviation of temperature from a reference state.

The representation for the inelastic strain rate with both back stress, Ω_{ij}, and drag stress, Z, can be expressed as

$$\dot{\varepsilon}_{pq}^{inel} = D_0 \exp\left[-\frac{A}{2}\left(\frac{Z^2}{3K_2}\right)^n \right] \frac{\left(S_{pq} - \Omega_{pq}\right)}{\sqrt{K_2}}.$$ (5)

The parameters A and n control the strain rate sensitivity, and D_0 is the limiting value for the inelastic strain rate. The difference between the deviatoric stress and deviatoric back stress, $S_{ij} - \Omega_{ij}$, is the over-stress which controls the onset of plastic slip (similar to the critically resolved shear stress) and the increase in resistance to plastic slip or strain hardening. The parameter K_2 is the second invariant if the over stress tensor, that is

$$K_2 = \frac{1}{2}(S_{ij} - \Omega_{ij})(S_{ij} - \Omega_{ij}).$$ (6)

The evolution equations for the back stress tensor and drag stress will be presented next. Since the maximum stress in a tensile test corresponds to a state of balance between strain hardening and dynamic recovery, a simple form for the back stress rate can be written as

$$\dot{\Omega}_{pq} = \frac{2}{3} f_1 \dot{\varepsilon}_{pq}^{inel} - f_1 \left(\frac{\Omega_{pq}}{\Omega_s}\right) \dot{\varepsilon}_{eff}^{inel} + f_2 \dot{S}_{pq},$$ (7)

where the first term models strain hardening and the second term represents dynamic recovery. The last term is added to control the yield stress and rate primary creep as a function of the initial load rate. In a constant strain rate tensile test the back stress rate approaches zero as the back stress approaches Ω_s, the saturated value for the back stress. The parameter $\Omega_s = \Omega_{max}$, a constant representing the maximum saturated back stress except during high temperature creep when static thermal recovery is present. The effective inelastic strain rate is defined similar to the second invariant of the inelastic strain rate tensor,

$$\dot{\varepsilon}_{eff}^{inel} = \frac{2}{3}\sqrt{\dot{\varepsilon}_{pq}^{inel}\dot{\varepsilon}_{pq}^{inel}}$$ (8)

The parameters f_1 and f_2 control the strain at which the strain hardening saturates and the initial yield stress, respectively.

The drag stress is a scalar quantity introduced to model cyclic hardening and softening. In most models cyclic hardening is taken as a function of the accumulated inelastic strain or the accumulated inelastic work, W^{inel}. The accumulated inelastic work has been used successfully for several materials because it is more sensitive for strain rate dependent materials, however the accumulated inelastic strain appears to be a little more convenient numerically. Thus, a representation for drag stress without thermal recovery is

$$Z = Z_1 + (Z_0 - Z_1) \exp(-mW^{inel}).$$ (9)

The saturated value of the drag stress is Z_1, the rate for cyclic hardening or softening is controlled by m, and $Z(0) = Z_0$ is the initial condition. Finally since the drag stress, Z, is a function of the accumulated inelastic work, changes in Z are expected to be small during small strain monotonic tensile and creep loading when compared to cyclic loading.

Next let us develop a representation for creep. To begin, Cadek (1987) has shown from theoretical derivation that back stress is present during creep, and the net force assisting the motion of dislocations is the over stress. However, in creep deformation the maximum value of the back stress is less than that found in tensile tests, Ω_{max}, due to a different deformation mechanism. There are two reasons for the lower back stress in creep. First, creep is driven by thermal vibrations and assisted by the presence of stress. Second, static thermal recovery results in a reduction of the number of dislocations over a long period of time. The steady state value of the back stress during creep can be plotted against the creep and saturated tensile stress as shown in Figure 3 to determine the effect of the different deformation mechanisms. For several alloys the following simple trilinear relationship has been found adequate

$$\Omega_{sat} = \begin{cases} \Omega_{max} & \text{for } \sigma > \sigma_1 \\ c_1 + c_2\sigma & \text{for } \sigma_2 > \sigma > \sigma_1, \\ c_3 & \text{for } \sigma < \sigma_2 \end{cases} \tag{10}$$

where c_1 and c_2 are constants, and c_3 is arbitrary small value used for numerical stability. Since the balance between strain hardening and recovery is different during creep and tensile loading, let us assume that the saturated back stress decays from an initial value $\Omega_s(0) = \Omega_{max}$ to the correct saturated value Ω_{sat} during long term creep, that is

$$\dot{\Omega}_s(t) = A \left(\frac{\sigma}{\sigma_o}\right)^p (\Omega_{sat} - \Omega_s). \tag{11}$$

During high rate tensile tests the stress increases fast, so the is no time for Ω_s to change from Ω_{max}. During creep the creep stress never gets as high as σ_1. The creep coefficient A and exponent p control the time required to reach secondary creep, and σ_o is for dimensional homogeneity.

4 MODIFICATIONS FOR MULTIAXIAL HARDENING

The first step in developing a representation for multiaxial hardening is to establish the tension-torsion equations from the tensor equations given in Section 2. Let us define σ, τ and Ω as the uniaxial stress, shear stress, and uniaxial back stress, respectively, where $\Omega = 2/3\ \Omega_{11}$. Then $K_2 = 1/3\ (\sigma - \Omega)^2 + (\tau - \Omega_{12})^2$ and $\left(\dot{\varepsilon}_{eff}^{inel}\right)^2 = \left(\dot{\varepsilon}_{11}^{inel}\right)^2 - 4/3\ \left(\dot{\varepsilon}_{12}^{inel}\right)^2$ so that the axial and shear components of Equations (5) and (7) are fully defined. The accumulated inelastic work in the drag stress equation is defined by the integral of $\dot{W}^{inel} = \sigma\ \dot{\varepsilon}_{11} + 2\tau\ \dot{\varepsilon}_{12}$.

The above equations have no facility to model the extra hardening observed in nonproportional loading. In general, there is a need to model extra hardening since it is present in most metals except at very high temperatures. In order to achieve this goal two changes are necessary. First a term must be added to the drag stress equation to increase its saturated value if nonproportional loading exists; and, second a method must be developed to sense the presence of nonproportional loading. Since the maximum amount of extra hardening occurs in 90 degree out-of-phase loading and there is no extra hardening during in-phase (proportional) loading, a measure of the "phase angle" can be used as a device to sense nonproportional loading. For strain rate controlled testing, the loading is proportional if the strain and strain rate are parallel, (Lindholm et. al., 1984), and nonproportional if they are not parallel, thus

$$\varepsilon_{pq}\ \dot{\varepsilon}_{pq} = \sqrt{\varepsilon_{st}\ \varepsilon_{st}}\ \sqrt{\dot{\varepsilon}_{nm}\ \dot{\varepsilon}_{nm}}\ \cos\theta \tag{12}$$

where $\cos\theta$ is the angle between the two vectors. A simple method to add extra hardening to the drag stress equation is to replace the saturated value Z_1 by $(1 + \alpha)\ Z_1$ in Equation (9), where α controls the amount of extra hardening. The parameter α can be written as

$$\alpha = \alpha_m \left(1 - |\cos\theta|\right) \tag{13}$$

where α_m is a scale factor that gives the maximum amount of extra hardening.

An evaluation of the extra hardening developed by these equations is given in Figures 4 and 5. It can be seen in Figure 4 that tensile and proportional loading do not produce any extra hardening, but the 90 degree out-of-phase loading does produce extra hardening. Figure 5 shows

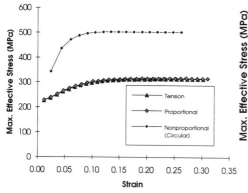

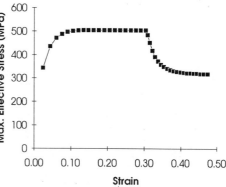

Figure 4 Cyclic hardening for three loading conditions ($\alpha_m=1$).

Figure 5 Predicted cyclic response when nonproportional loading is followed by proportional loading.

that if nonproportional loading is followed by proportional loading, the extra hardening can be removed. This assumption seems reasonable since the dislocation microstructure would be expected to evolve into that characteristic of the current loading condition. Figure 6 shows application of the equations to Rene' 80 at 982 °C in nonproportional loading with and without the extra hardening modification. It can be seen that the extra hardening parameters are sufficient to improve the accuracy of the model results.

5 MODIFICATIONS FOR THERMOMECHANICAL FATIGUE

A number of in-phase and out-of-phase thermomechanical fatigue tests were run on Rene` 80 in the temperature range between 649°C and 1093°C. The isothermal tests were at 538°C, 649°C, 832°C, and 982°C, and the model parameters were calibrated at these temperatures. Direct application of the model given in Section 3 by extrapolation and interpolation of the material parameters was not fully successful. The stress range and shape of the curves did not match the experimental data as well as desired (D. C. Stouffer and V. G. Ramaswamy, 1988). This resulted from three major effects. First at about 700°C there was a rather abrupt change in the deformation mechanism from planer slip to slip and climb. This made interpolation difficult. Second, extrapolation to 1093°C from 982°C was not possible because there was an unexpected increase in recovery above about 1000°C. Third, cycling between temperatures above and below 700°C introduced a thermal history effect that could not be determined by isothermal testing alone.

There were three modifications to the equations to correct for the above effects. To begin, the flow equation is very similar to the Arrhenius equation from physical metallurgy for steady state creep as a function of stress and temperature. Thus temperature was added to the flow equation, and Equation (5) is replaced by

$$\dot{\varepsilon}_{pq}^{inel} = D_o \exp\left[-\left(\frac{Z^2}{3\,T^2\,K_2}\right)^n\right] \frac{\left(S_{pq} - \Omega_{pq}\right)}{\sqrt{K_2}} \tag{14}$$

where the parameter A/2 has been eliminated and the values of the parameters in the drag stress, Z, are different. Temperature is on the Kelvin scale, and the drag stress constants were re-evaluated and verified at each of the isothermal temperatures. Since all the experimental results are for uniaxial thermomechanical loading, the uniaxial form of Equation (14) is

$$\dot{\varepsilon}^{inel} = \frac{2}{\sqrt{3}}\,D_o \exp\left[-\left(\frac{Z}{T|\sigma-\Omega|}\right)^{2n}\right] \frac{(\sigma - \Omega)}{|\sigma - \Omega|} \tag{15}$$

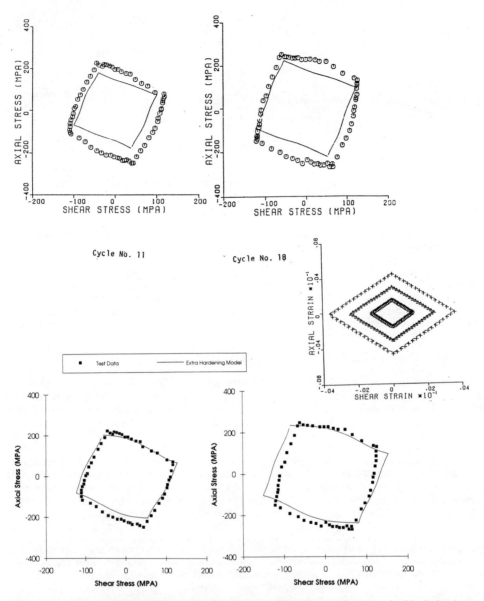

Figure 6. Comparison of experimental and models results without (above) and with (below) extra hardening for Rene` 80 at 982 °C loaded as show in the inset.

where σ and Ω are the uniaxial stress and back stress, respectively.

The Back Stress and Drag Stress Equations, (7) and (9), respectively, remain the same. Further it was observed that the thermomechanical fatigue loops saturate in a few cycles and there does not appear to be a significant long term change in the saturated response. Thus long time static recovery term does not appear necessary. Next note that the parameter Z_1 is determined to characterize the saturated equilibrium dislocation structure (balance between hardening and recovery) at each temperature. In addition the exponent m and coefficients Z_0 and Z_1 are updated during thermomechanical cycling. For example, let us assume that the deformation history starts at some temperature T_0 and the changes rapidly to T_1 before the onset of inelastic deformation, and then remains at T_1 for the duration of the experiment. The parameters m, Z_0 and Z_1 in the drag stress equation will be updated from their values at T_0 to their corresponding values at T_1 and the correct inelastic response at T_1 will determined.

Consider next the Thermal Recovery Equation, (11). Let us start by applying the same temperature test as described for the drag stress. In this case the change of temperature from T_0 to T_1 after the test starts is not recognized by the Equation (11) since $\Omega_s(0) = \Omega_{max}$ at $T(0)=T_0$. Thus, Equation (11) can not give the correct result. Since the rate of thermal cycling is very slow (a typical cyclic requires 6 to 10 minutes) the thermal and stress rates can be neglected and Equation (11) can be integrated to obtain

$$\Omega_s = \Omega_{sat}(T) + \left[\Omega_{max}(T) - \Omega_{sat}(T)\right] \exp\left[-A(T)\left(\frac{\sigma}{\sigma_o}\right)^p t_r\right] \tag{16}$$

The time t_r is the accumulated (elapsed) time the material is exposed to a temperatures above 700°C, the temperature at which diffusion and climb become important. The exponential controls the amount of recovery during the time t_r, at the stress level σ.

In general time is not considered as an acceptable variable for a constitutive equation. Usually time appears in an evolution equation that must be integrated as part of the solution algorithm. In these cases an elapsed time variable must be used to satisfy objectivity. This results in an integral with poor numerical properties. However, Equation (16) is not a differential equation and does not have to be integrated, so this is not an issue.

The first evaluation of the new equations was to examine the response to a constant temperature rate between the isothermal temperatures 538 °C, 760 °C, 871 °C, and 982 °C by extrapolation. As a result of the exercise it was found that the tensile response curves at constant temperatures between the four isothermal test temperatures were not even ordered when calculated by interpolation. This was because the parameters Z_0 and Z_1 in the drag stress equation and the strain rate sensitivity exponent , n, in the flow equation changed rapidly at about 700 °C (1050 °K). This corresponds to the temperature at which dislocation climb becomes important. As a result these three parameters were adjusted by trial and error until a smooth variation in tensile response with temperature was achieved. The final variation of Z_0, Z_1, and n in temperature are shown in Figure 7. The increase in drag stress with temperature occurs because climb adds or subtracts atoms to the dislocation line and it lies in several planes simultaneously increasing the stress to produce slip. The subsequent variation in tensile response to increasing and decreasing temperature rate is given in Figure 8. The jumps in stress during the temperature change result from the steep slope of the curves

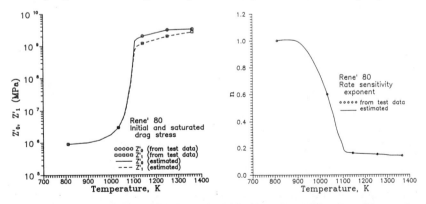

Figure 7 Variation of the parameters Z_0, Z_1, and n with temperature. The steep slopes at the change in deformation mechanism (approximately 1050 °K) made interpolation impossible.

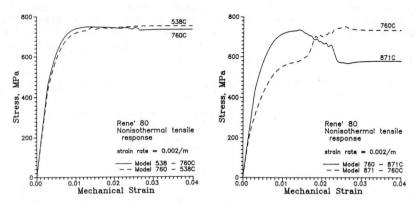

Figure 8 Variations in stress (worst case on right) due to increasing and decreasing temperatures changes between 760 °C and 871 °C.

in Figure 7, however this type of behavior was experimentally observed by Chan, 1988, for the nonisothermal tensile response of B1900+Hf between 538 °C and 760 °C.

Consider next the response at 1093 °C that can be determined by extrapolation. Figure 9 gives a comparison between the experimental and extrapolated response. It is clear that the experimental results are much softer (smaller stress range and larger inelastic strain range) than that predicted by the model. As a result material parameters were determined from experimental data at 1093 °C. The correlation between the experimental data and the model using the revised parameters is much better than the extrapolated parameters as shown in Figure 9. Thus extrapolation to high temperatures was not successful.

Let us next begin the task of modeling thermomechanical fatigue. A comparison between the first cycle of experimental and model results for an out-of-phase TMF history is shown in Figure 10. The accuracy of these results confirmed the modifications to the model as determined from the isothermal data was successful. Let us next examine the response to several cycles of thermomechanical loading as shown in Figure 11. The results show that the compressive straining at 1093 °C had little effect on the subsequent response at tensile response at 649 °C, however , the low temperatures tensile response did effect the subsequent high temperature properties. The predicted response is about 40 MPa below the experimental data and the shape of the calculated curve in the second cycle is not correct.. The material appears to be harder than that predicted by the model.

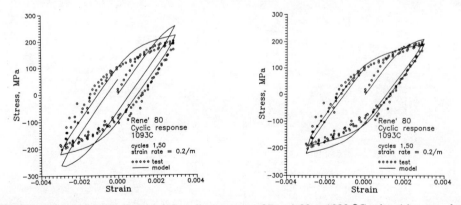

Figure 9 Experimental and calculated cyclic response of Rene` 80 at 1093 °C using (a) extrapolated parameters, and (b) using parameters determined from test data at 1093 °C.

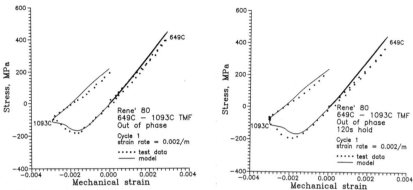

Figure 10 Experimental and calculated first cycle response of Rene' 80 without (left) and with at 120 second hold in compression.

The reason for this extra hardening can be explained by examining the deformation mechanisms that are active during the test. At the lower temperatures deformation is by planer slip, where as at the higher temperatures diffusion is much more important and the deformation involves dislocation climb and recovery. Deformation at some fixed temperature will produce a microstructure that is characteristic of the temperature. Changing temperatures will force the deformation to occur in a microstructure that is not characteristic of that temperature. In this case the inelastic deformation at 1093 °C occurs by slip and climb, and there will be a significant amount of recovery. The response on unloading from 1093 °C to almost 649 °C is nearly elastic. The subsequent inelastic deformation at 649 °C occurs in a relatively soft material by planer slip and without much recovery thereby increasing the dislocation density and hardness. The subsequent inelastic deformation at 1093 °C in the second cycle, occurs after elastic unloading in the microstructure created at 649 °C with an increased dislocation density. Thus the response in the second cycle at 1093 °C is harder than that observed in the first cycle.

A model was developed for the extra hardening from the following assumptions:
(1) The thermomechanical history resulted in extra hardening at the higher temperatures.
(2) Temperature variations must be accompanied by inelasticity to produce extra hardening.
(3) Isothermal deformation after thermomechanical cycling can erase the extra hardening.

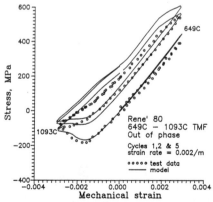

Figure 11 Experimental and calculated response to cycling between 649 and 1093 °C as determined from the isothermal parameters by interpolation.

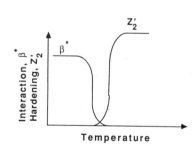

Figure 12 Definition of the extra hardening parameters β^* and Z_2

173

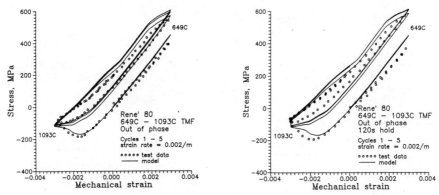

Figure 13 Experimental and calculated response for the first five cycles with (left) and without (right) a 120 second hold in compression.

Since the extra hardness is isotropic and it occurs in cyclic loading, the drag stress equation is modified to include an extra hardening term. ω as follows:

$$Z = Z_1 + (Z_0 - Z_1)\exp(-mW). + \omega \qquad (17)$$

where ω is defined by

$$\omega = Z_2 \left[1 - \exp\left(-\beta \; W^{inel}\right) \right]. \qquad (18)$$

The hardening parameter ω is expected to vanish in isothermal loading. The quantity β is an interaction variable defined to sense the presence thermal cycling, W^{inel} is the accumulated inelastic work during thermal cycling, and magnitude of the function Z_2 is the maximum amount of extra hardening that is developed at the higher temperatures during thermomechanical cycling. The parameter β is defined as

$$\dot{\beta} = m_2 \left(\beta^* - \beta\right) \dot{W}^{inel} \qquad (19)$$

so that it has the following properties:

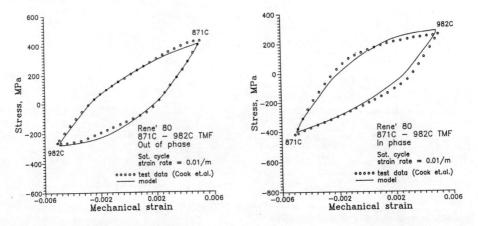

Figure 14 Experimental and model results to out-of-phase (left) and in-phase thermomechanical cycling between 982 °C and 871 °C.

174

$$\beta\left(\dot{w}^{inel} = 0\right) = 0$$
$$\beta \neq 0 \quad \text{when} \quad \dot{w}^{inel} \neq 0 \quad \text{and} \quad T < T_c = 700\ ^\circ C \qquad (20)$$
$$\beta = 0 \quad \text{when} \quad \dot{w}^{inel} \neq 0 \quad \text{and} \quad T > T_c = 700\ ^\circ C$$

Both Z_2 and β^* are stiff exponential functions representing essentially step up and step down functions, respectively, as shown in Figure 12. During thermal cycling at the low temperatures $\beta^* \neq 0$, $\beta \Rightarrow \beta^*$ as the deformation progresses and $Z_2 = \omega = 0$. However, during high temperature deformation $\beta^* = 0$, $\beta \Rightarrow 0$, and $Z_2 \neq 0$ and $\omega \neq 0$. Thus the three assumptions stated above are satisfied.

The results of this modification is given in Figures 13, 14, and 15. Shown is a comparison between the experimental and model results for the first five cycles of an out-of-phase test between 649 $^\circ$C and 1093 $^\circ$C, and a similar test except with a 120 second hold at 1093 $^\circ$C. In both figures the experimental and model results were fully saturated. Experimental and model predictions for in-phase and out-of phase thermomechanical cycling between 982 $^\circ$C and 871 $^\circ$C is given in Figure 14. Also shown in Figure 15 are the maximum and minimum stress during mechanical cycling at five different temperature sequentially. In all cases there is reasonable agreement between the experimental results and the model predictions.

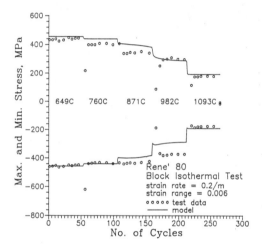

Figure 15 Experimental and predicted results for the maximum and minimum stress during cycling at five temperatures sequentially.

6 CONCLUSIONS AND OBSERVATIONS

These results are a step in understanding and modeling two rather complicated phenomena. In both cases there is some motivation for additional experiments. In multiaxial fatigue it would be interesting to find out if the extra hardening is erased by proportional loading after nonproportional loading. A experimental thermomechanical program should include several tests at and across the temperature range where climb becomes important. These could include isothermal tests, simple temperature rate tests, and isothermal loading after thermomechanical loading. Microstructural examination would also be very helpful.

Modeling of thermomechanical fatigue needs more attention. Time is not a very good variable for modeling. The combined effect of multiaxial thermomechanical cycling also needs to be examined both experimentally and analytically. This is the type of loading that many structures actually experience.

REFERENCES

Bannantine, J. A., Comer, J., L., and Handrock J., L., 1990, *Fundamentals of Metal Fatigue Analysis,* Prentice Hall Inc., Englewood Cliffs, New Jersey.

Cadek J., 1987, The Back Stress Concept in Power Law Creep of Metals, *Mat. SCI. Eng..,* Vol. 94. pp. 79-92.

Chan K. S.. and Page R. A., 1988, "Inelastic Deformation and Dislocation Structure of Nickel Alloy: Effects of Deformation and Thermal Histories", *Metall. Trans.,* 19A:1477-2486.

Courtney, T. H., 1990, *Mechanical Behavior of Materials,* Mc Graw Hill.

Dieter, G. E., 1986, *Mechanical Metallurgy,* Mc Graw Hill, 3rd. ed..

Lindholm, U. S.., Chan, K. S.., Bodner, S. R.., Weber, R. M.., Walker, K. P.., and Cassenti, B. N.., 1984, "Constitutive Modeling for Isotropic Materials," NASA CR174718.

Manson, S. S. and Halford G. R., 1981, "Practical Implementation of the Double Linear Damage Rule and Damage Accumulation Curve Approach for Cumulative Damage", *Int. J. Fracture,* Vol. 17.2, pp. 169-172.

McDowell, D., S., 1983, "Transient Nonproportional Cyclic Plasticity," University of Illinois at Urbana Champaign, Report No. 107, UILU-Eng.-83- 4003.

Morrow, J., 1968, *Advances in Engineering,* Vol. 4, Society of Automotive Engineers, Warrendale Pa, Sec 3.2, pp. 3-36.

Ramaswamy, V. G.., Stouffer, D. C.. and Laflen, J. H.., 1990, "A Unified Constitutive Model for the Inelastic Uniaxial Response of Rene'80 at Temperatures Between 538C and 982C" ASME *Journal of Engineering Materials and Technology,* Vol. 112.

Ramaswamy, V. G.., Van Stone, R. H.., Dame, L. T.. and Laflen, J. H., 1984, "Constitutive Modeling for Isotropic Materials," NASA Conference Publication 2339.

Ramaswamy, V. G.. , 1986, "A Constitutive Model for the Inelastic Multiaxial Cyclic Response of a Nickel Base Superalloy Rene'80, CR NAS3-23927.

Sheh, M., Y., and Stouffer, D., C., 1990, "A Crystallographic Model for Tensile and Fatigue Response of Rene' N4 at 982C.", ASME *Journal of Applied Mechanics,* V57, pp 25-31.

Smith, K., N., Watson, P., and Topper, T., H., 1970, "A Stress-Strain Function for the Fatigue of Metals," *J. Mater.,* Vol.5.4, pp. 767-778.

Stouffer, D., C., and Ramaswamy, V., G., 1988, "A Constitutive Model for the Uniaxial, Multiaxial, and Thermomechanical Response of Rene` 80 Between 538C and 1093C", *PVP.* Vol. 153, D. Hui and T., Kozik Eds., pp. 101-107.

Stouffer, D. C.., Ramaswamy, V. G.., Laflen, J. H.., Van Stone R. H.., and Williams, R., 1990, "A Constitutive Model for the Inelastic Multiaxial Response of Rene'80 at 871C and 982C", ASME *Journal of Engineering Materials and Technology,* Vol. 112

Wheeler, O., E., 1972, "Spectrum Loading and Crack Growth", ASME *J. Basic Engineering,* Vol. D94.1, pp. 181-186.

Willenborg, J., Engle, R., M., and Wood, H., A., 1971, "A Crack Growth Retardation Model Using an Effective Stress Concept," AFFDL-TM-71-1-FBR.

AD-Vol. 36, Fatigue and Fracture of Aerospace Structural Materials
ASME 1993

ELEVATED TEMPERATURE CRACK PROPAGATION

Thomas W. Orange
Fatigue and Fracture Branch
NASA Lewis Research Center
Cleveland, Ohio

ABSTRACT

This paper is a summary of two NASA Contracts on high temperature fatigue crack propagation in metals. The first, begun in 1980, evaluated the ability of then-current nonlinear fracture parameters to correlate crack propagation. Hastelloy-X specimens were tested under isothermal and thermomechanical cycling at temperatures up to 980°C (1800°F). The most successful correlating parameter was the crack tip opening displacement derived from the J-integral.

The second program was more extensive. It evaluated the ability of several path-independent integrals to correlate crack propagation behavior. Eight integrals were first evaluated from a theoretical standpoint. Inconel 718 specimens were tested under isothermal, thermomechanical, temperature gradient, and creep conditions at temperatures up to 650°C (1200°F). The integrals formulated by Blackburn and by Kishimoto were able to correlate the data reasonably well under all test conditions.

INTRODUCTION

This paper reviews and summarizes two NASA Contracts on high temperature fatigue crack propagation in metals, with application to aircraft turbine engine hot-section components. The first contract was begun in 1980. Its objectives were (1) to determine those engine hot-section components for which linear elastic fracture mechanics (LEFM) analysis methods are inadequate, and (2) to evaluate the ability of then-current nonlinear fracture parameters to correlate crack propagation. Hastelloy-X specimens were tested under isothermal and thermo-mechanical cycling. Maximum test temperature was 980°C (1800°F) and peak loads were well beyond yield.

The second contract, begun in 1983 as a part of the NASA HOST (HOt Section Technology) Program, was more extensive. This contract consisted of a basic program and an option program. The basic program evaluated the ability of several path-independent integrals to correlate time-independent crack propagation. Eight integrals were first evaluated from a theoretical standpoint. Computer simulations of the tests to follow were run to determine whether the integrals were indeed path-independent under the proposed test conditions. Inconel 718 specimens were tested under isothermal, thermomechanical, and temperature gradient conditions. Maximum test temperature was 650°C (1200°F) and peak loads were also well beyond yield.

The option program was incorporated into NASA's Earth-to-Orbit Propulsion Technology Program. It extended the previous studies to time-dependent behavior using similar procedures. Rate versions of selected integrals were evaluated from a theoretical standpoint. Computer simulations of the tests to follow were run. Specimens were tested under static loading, strain cycling with hold time, and thermomechanical loading with hold time. Also, data from the first contract were re-analyzed to see if a better correlation could be obtained using path-independent integrals.

<u>FIRST CONTRACT</u>

<u>Approach</u>

The first contract was awarded to Pratt & Whitney Aircraft (Commercial Products Div.). The Principal Investigator was G.J. Meyers. First, an "engine survey" was conducted to determine those areas of a turbine engine hot section where LEFM methods are not adequate to predict crack propagation rates. The survey was based on experience with the JT9D engine, an advanced commercial turbofan.

Isothermal and thermomechanical fatigue (TMF) cyclic tests were conducted. The test matrix (strain range, temperature) is shown in Fig. 1 and the TMF cycles in Fig. 2. The "faithful cycle" is a simplification of results from a finite element analysis of a combustor liner louver. Most of the TMF tests were out-of-phase (OP), but one in-phase (IP) test and one "faithful cycle" test were run.

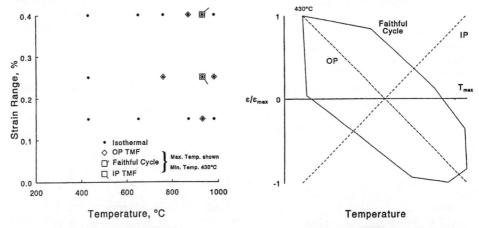

Fig. 1 - Hastelloy-X test matrix.   Fig. 2 - Hastelloy-X TMF cycles.

Tubular specimens with circumferential cracks (Fig. 3) were made of Hastelloy-X, a typical combustor liner material. Isothermal tests were run in an electric furnace. TMF specimens were heated by induction coils and cooled by forced air. Most tests were conducted with zero mean strain. Displacements were measured over the gage length only (i.e., crack mouth displacements were not measured). Crack lengths were measured optically.

The parameters that were compared were the stress intensity factor, the strain intensity factor, the J-integral, the crack tip opening displacement (CTOD), and Tomkins' (1975) model. The stress and strain intensity factors for the cracked tubular specimen were based on an analysis by Erdogan and Ratwani (1970). The same stress intensity factor was used to compute the elastic portion of the J-integral, and the plastic portion used an expression by Shih and Hutchinson (1976) for a small crack in a flat plate. A simplified crack opening displacement (CTOD) was calculated using the Dugdale (1960) model. For some of the isothermal tests, CTOD was also calculated from the J-integral and the strain hardening coefficient.

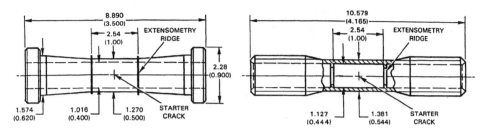

Fig. 3 - Hastelloy-X tubular test specimens. Dimensions in cm (in.).

Results

The results of this contract were reported by Meyers (1982). Portions of this work have also been reported by Jordan and Meyers (1986, 1989).

Results of the engine survey indicated that cyclic nonlinear material behavior seldom occurs in turbine disks, seals, spacers, and cases. For these components, LEFM analysis is adequate. Life prediction for blades and vanes is generally based on crack initiation life, with a correction factor based on engine experience. Since a 3D strain and temperature analysis of a blade or vane would be difficult and costly, nonlinear analysis would only be used for investigating serious safety concerns. The combustor liner, however, is fairly simple to analyze and is a high-maintenance-cost item. Nonlinear analysis would be most useful in this case.

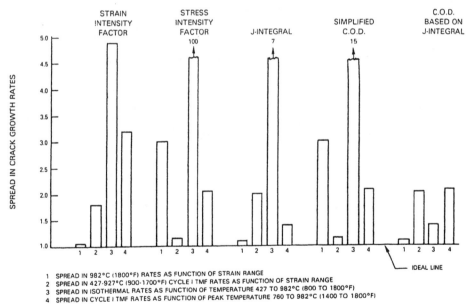

1 SPREAD IN 982°C (1800°F) RATES AS FUNCTION OF STRAIN RANGE
2 SPREAD IN 427-927°C (900-1700°F) CYCLE I TMF RATES AS FUNCTION OF STRAIN RANGE
3 SPREAD IN ISOTHERMAL RATES AS FUNCTION OF TEMPERATURE 427 TO 982°C (800 TO 1800°F)
4 SPREAD IN CYCLE I TMF RATES AS FUNCTION OF PEAK TEMPERATURE 760 TO 982°C (1400 TO 1800°F)

Fig. 4 - Spread of crack growth rates for various fracture parameters.

Experimental results are not totally consistent, as can be seen in Fig. 4. Surprisingly, the stress intensity factor collapsed the OP TMF growth rates as a function of strain range (bar 2) rather well. But, as expected, it did not do well overall. The CTOD based on the J-integral was the best overall. The other methods all failed to collapse the isothermal growth rates as a function of temperature (bar 3). Three forms of Tomkins' model were applied to some of the isothermal data, but the correlation was no better than for the strain intensity factor and it was not carried further. An attempt was made to predict the TMF behavior from the isothermal

179

data by integrating the instantaneous growth rate (as a function of temperature) over the tensile portion of the cycle. The prediction was good for an IP TMF test but only fair for and OP TMF and the "faithful cycle" test.

After the conclusion of this contract, it was apparent that a more fundamental approach to elevated temperature crack propagation was needed. There were a number of J-like path-independent integrals which looked promising but which had not been critically evaluated. Also, the phenomenon of crack closure (Elber, 1971) should be considered in any analysis.

SECOND CONTRACT, BASIC PROGRAM

Approach

The second contract was awarded to GE Aircraft Engines. The Principal Investigator for the basic program was J.H. Laflen. First, eight path-independent integrals were evaluated from a theoretical standpoint. Those included were denoted J (Rice, 1968), J_W (Wilson and Yu, 1979), J_G (Gurtin, 1979), J_θ (Ainsworth et al., 1978), J* (Blackburn, 1972), $\hat{J}$ (Kishimoto et al., 1980), and the ΔT_p and ΔT_p^* integrals by Atluri et al. (1983). The Atluri integrals are based on the incremental theory of plasticity, the rest on the deformation theory.

The path-independence of the integrals under cyclic loading, thermomechanical loading, and cyclic loading with a temperature gradient was first investigated analytically. The model used was an edge-cracked strip under uniform end displacements. Analyses were made using CYANIDE, a General Electric nonlinear finite element program. Gap elements were used to model crack closure. A postprocessor was written to calculate each integral along several integration paths. The four integrals were all found to be acceptably path-independent.

For testing, an analog material was desired which can be tested at lower temperatures (to ease instrumentation problems) while retaining many of the important characteristics of a combustor liner material. These characteristics include significant variation in elastic modulus, large changes in short-time creep rates, and absence of phase transformations within the test temperature range. Inconel 718 in the temperature range 430-650°C (800-1200°F) was selected.

To meet the program requirements, special consideration was given to the type of test specimen to be used. The specimen should be suitable for isothermal and TMF testing under tension-compression loading and should permit the establishment of a temperature gradient in the direction of crack propagation. Displacements at the load point and at the crack mouth must be measured. For these reasons, a buttonhead single-edge-crack specimen was chosen, and is shown in Fig. 5.

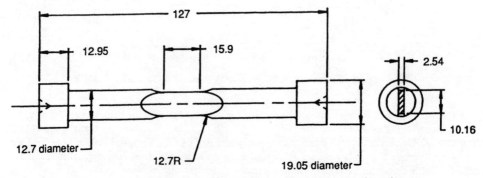

Fig. 5 - Buttonhead single edge crack specimen. Dimensions in cm.

Elastic-plastic 3D finite element analyses of the specimen were performed over the range of crack lengths to be tested. They showed that the displacements at the ends of the gage section were not constant but, rather, varied linearly. This was confirmed by experimental measurement. Thus it was necessary to model only the gage section, with end displacements varying linearly.

The instrumentation that was used is shown in Fig. 6. An extensometer on the specimen centerline is used for strain control. The back-face extensometer, combined with the control extensometer, is used to determine the linear displacement gradient. The crack-mouth opening displacement (CMOD) gage is used in determining crack closure and the DC potential drop method to determine crack length. These measurements were recorded continuously using a computerized data acquisition system.

Isothermal, IP and OP TMF, and temperature gradient cyclic tests were conducted over the range 430-650°C (800-1200°F). The test matrix is shown in Fig. 7. The specimens were induction heated to establish and maintain the desired temperature. For the temperature gradient tests, cool air was blown on the back face. The resulting gradient was approximately

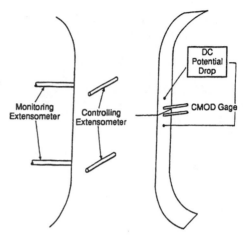

Fig. 6 - Displacement instrumentation.

trilinear, as shown in Fig. 8, and was reasonably uniform along the length of the specimen. The crack was grown under cyclic loading from an initial length of about 0.5 mm to a final length of 8-12 mm. Cyclic rate for all tests was 0.6 cpm.

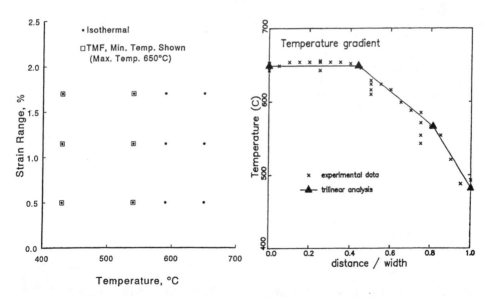

Fig. 7 - Test matrix for basic program.   Fig. 8 - Trilinear temperature gradient.

The path-independent integrals were calculated from the test measurements as follows. First the load-CMOD loops were analyzed to determine the crack opening and closing loads. At selected crack lengths, the control and backface displacements were used as inputs to the CYANIDE program where the integrals were calculated.

Results

The results of the basic program have been reported by Kim (1985), Yau et al. (1985), Malik et al. (1987), and Kim et al. (1990a). Portions of this work have also been reported by Kim and Orange (1988), Kim et al. (1990b), and Kim and Van Stone (1990).

The evaluation of the path-independent integrals was reported in detail by Kim (1985) and by Kim and Orange (1988). The results are summarized in Table 1. For engineering application, a path-independent integral must satisfy three requirements. First, the integral must be calculable without great difficulty and must be reasonably path-independent. Second, it must be possible to determine the value of the integral from measurements made on a test specimen. Third, the integral must consolidate various types of crack propagation data into a single curve.

Table 1. - Evaluation of Path-Independent Integrals

Integral	Non-Proportional Cyclic Loading	Temperature Gradient	Material Inhomogeneity	Elastic-Plastic Strains
J	No	No	No	Yes
J_θ	No	Yes	No	Yes
J_W	No	Yes	No	No
J_G	No	Yes	No	No
J^*	Yes	Yes	Yes	Yes
$J^\wedge$	Yes	Yes	Yes	Yes
ΔT_p	Yes	Yes	Yes	Yes
ΔT_p^*	Yes	Yes	Yes	Yes

Note that six integrals are comprised of a line integral plus an area integral. The first four are not suitable for the situations characteristic of turbine engine hot sections. The last four integrals seemed the most promising and were selected for further evaluation. But note that three of the four cannot be measured experimentally. That is, since they cannot be expressed as a rate of change of potential energy they cannot be evaluated from load-point displacement measurements. Nor can they be evaluated by measuring strains around a contour remote from the crack tip, as was done by King and Herrmann (1981). A full-field stress-strain analysis would be required.

The final report for the basic program was given by Kim et al. (1988). Crack closure was found not to be a significant factor in these experiments. The crack opening load was very near the minimum load. Iyyer and Dowling (1985) found a similar trend in tests of 4340 steel with small surface cracks under fully reversed loading at high plastic strain levels.

As expected, the elastic stress intensity factor did not correlate the isothermal crack growth rates as a function of cyclic strain level. Data for 540°C are shown in Fig. 9. But all four integrals correlated these tests quite well. That is, crack growth rate data for all three strain levels could be fitted to the same straight line. An example (for the Atluri integral ΔT_p^*) is shown in Fig. 10. The three other integrals were equally successful.

The TMF and temperature-gradient tests were completed under this program, but their analysis was deferred to the option program and were reported by Kim and Van Stone (1992). However, they will be discussed here for consistency. Three integrals (Atluri's ΔT_p^*, Blackburn's J^*, and Kishimoto's $\hat{J}$) correlated the TMF data quite well. An example (for Kishimoto's $\hat{J}$) is shown in Fig. 11. Atluri's ΔT_p integral did not do quite as well.

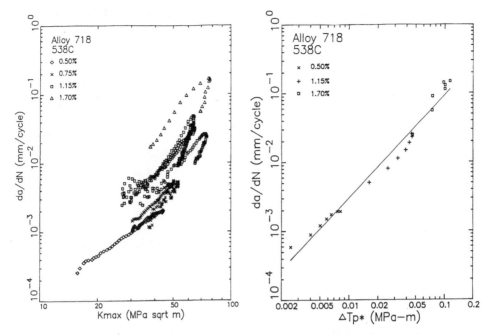

Fig. 9 - Isothermal crack growth rate as a function of ΔK_{max}.

Fig. 10 - Isothermal crack growth rate as a function of ΔT_p^*.

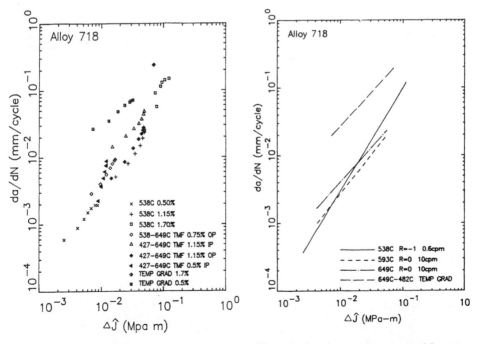

Fig. 11 - Isothermal, TMF, and temp. gradient crack growth rate as a function of $\Delta \hat{J}$.

Fig. 12 - Crack growth rate trend lines as a function of ΔJ^* for isothermal and temp. gradient tests.

Note in Fig. 11 that the temperature-gradient data for the two strain ranges align well with one another but not with the rest of the data. The gradient crack propagation data was taken in the constant-temperature (650°C) portion of the specimen (the integration path, however, did contain the gradients). Unfortunately the 650° isothermal data were not analyzed. But in Fig. 12 the trend line for the 650-480° gradient tests can be compared with 650° isothermal data from the hold-time series (zero hold time), which were run at a much faster rate. The curves are nearly parallel. The difference may be due to time-dependent crack growth occurring during the slower gradient tests.

SECOND CONTRACT, OPTION PROGRAM

Approach

The Principal Investigator for the option program was R.H. Van Stone. This program extended the previous studies to time-dependent behavior using similar procedures. First, the rate versions of four integrals were evaluated from a theoretical standpoint. These were the C^* integral and the rate versions of J^* (Blackburn, 1972), $\hat{J}$ (Kishimoto et al., 1980), and ΔT_p^* (Atluri et al., 1983). C^* is the rate version of the J integral, as formulated by Goldman and Hutchinson (1975) and so named by Landes and Begley (1976).

The rate integrals were first evaluated from a theoretical standpoint. Next the computational path-independence was studied using the edge-cracked strip model from the basic problem. The cases used were uniform end displacement, uniform stress, and uniform stress with a temperature gradient (decreasing in the direction of crack advance).

Static loading tests were run at 590°C and 650°C with strain levels of 0.4, 0.75, and 1.15%. Two tests with a constant load of 700 MPa were also run at the same temperatures. Cyclic load tests (with hold time) were chosen to represent a simple combination of time-independent and time-dependent loading. The test matrix for these tests is shown in Table 2. TMF tests (with hold time) were chosen to represent a more complicated combination. Strain levels were 0.40% IP and both IP and OP at 1.15%. For all three conditions, tests were run with no hold time, with 300 s hold at maximum load, and with 300 s hold at zero load. The instrumentation and method of calculating the integrals from test measurements were the same as in the basic program.

Table 2. - Test Matrix for Cyclic Load with Hold Time

Temp., °C	Strain (%) at Hold Time		
	0 s	30 s	300 s
540			0.4, 1.15
590	0.4, 1.15	0.4, 1.15	0.4, 1.15
650		0.4, 1.00	0.4, 1.15

In addition, the experimental results from the first contract (Meyers, 1982) were re-analyzed to see if the path-independent integrals would help consolidate these data. The Hastelloy-X tubular specimen was modelled as a center-crack flat plate with uniform end displacements. Bending of the specimen due to the presence of the crack was assumed negligible. Finite element analyses required only minor modifications to the mesh used for the single edge-crack specimen. Additional data required for the analyses were obtained for Hastelloy-X from published sources.

The results of the option program were reported by Kim and Van Stone (1992). The evaluation of the path-independent rate integrals is summarized in Table 3. The C^* integral, as expected, is not suitable but the other three hold promise.

Table 3. - Evaluation of Path-Independent Rate Integrals

Integral	Non-Proportional Cyclic Loading	Creep Deformation	Temperature Gradient	Elastic-Plastic Strains
J^*	Yes	Yes	Yes	Yes
$J^\wedge$	Yes	Yes	Yes	Yes
ΔT_p^*	Yes	Yes	Yes	Yes
C^*	No	Steady State Only	No	No

The rate versions of both the J^* and the $\hat{J}$ integrals collapsed the static loading tests (constant load as well as constant strain) equally well. An example, for the rate version of the $\hat{J}$ integral at 650°C, is shown in Fig. 13. The rate version of Atluri's ΔT_p^* did not correlate well at all.

The cyclic load with hold time was first analyzed with a simple superposition method,

$$(da/dn)_{hold\ time} = (da/dn)_{cyclic} + t_h \cdot (da/dt)$$

where t_h is the hold time. The results for the J^* integral and its rate version (Fig. 14) show fairly good correlation for two strain ranges. Better correlation was obtained using a modified superposition model,

$$(da/dn)_{hold\ time} = (da/dn)_{cyclic} + \beta \cdot t_h \cdot (da/dt)$$

where $0 \le \beta \le 1$ and β is a function of maximum strain, hold time, and temperature. When the constant β was determined to minimize the standard deviation between the predicted and experimental crack growth data for individual specimens, a much better correlation resulted (Fig. 15). The constant β is a function of maximum strain, hold time, and temperature, but more tests would be required to define it fully.

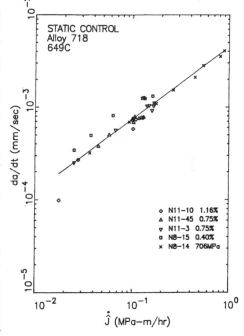

Fig. 13 - Isothermal time rate of crack growth as a function of $\Delta \hat{J}$

The TMF with hold time data were not analyzed using path-independent integrals. However, the results were qualitatively similar to those obtained separately for TMF and for static loading. Thus the TMF with hold time data could probably be modeled successfully with path-independent integrals.

The re-analysis of the Hastelloy-X data from the first contract was fairly successful. The J^*, $\hat{J}$, and ΔT_p^* integrals correlated the isothermal data well. The ΔT_p integral was not as successful and seemed to worsen with increasing temperature. An example for the ΔT_p^* integral at 930°C is given in Fig. 16. The same three integrals also correlated the 430-870°C out-of-phase (OP) TMF data well, and an example (for the ΔT_p^* integral) is shown in Fig. 17. However, the

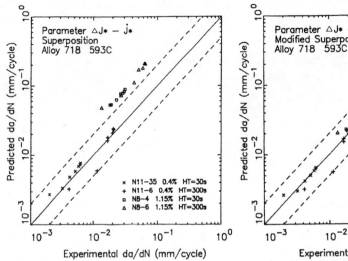

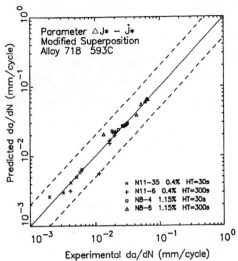

Fig. 14 - Cyclic loading with hold time; crack growth rate prediction by simple superposition using $\Delta J*$.

Fig. 15 - Cyclic loading with hold time; crack growth rate prediction by modified superposition using $\Delta J*$.

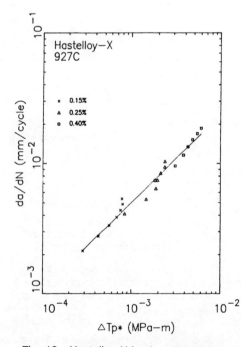

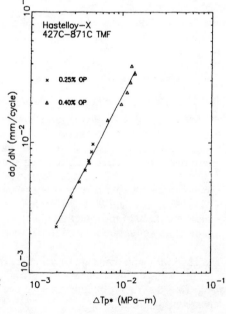

Fig. 16 - Hastelloy-X isothermal crack growth rate as a function of ΔT_p*.

Fig. 17 - Hastelloy-X TMF crack growth rate as a function of ΔT_p*.

correlation for all three integrals worsened as the maximum cycle temperature increased. This is probably due to creep crack propagation, which increases with temperature and was not accounted for in the analyses.

SUMMARY

The first contract showed that there was a need for methods for correlating fatigue crack growth in turbine engine hot sections. It also showed that linear elastic fracture mechanics methods are not adequate and that simple nonlinear methods are but a slight improvement.

Based on expectations at the beginning of the second contract, the biggest disappointment was in the analytical and experimental effort needed to calculate the path-independent integrals. Another was the length of time needed to complete the second contract, even though most of the delays were unavoidable. A third was the fact that we were usable to complete the analyses for the TMF-holdtime and some of the isothermal tests. And, in hindsight, the test specimen was probably not optimum for the temperature gradient tests.

The greatest success was the performance of the path-independent integrals. Four of the integrals (Atluri's ΔT_p and $\Delta T_p{}^*$, Blackburn's J^*, and Kishimoto's $\hat{J}$) were all able to correlate the isothermal data. Three ($\Delta T_p{}^*$, J^*, and $\hat{J}$) were able to correlate the TMF data. Two (J^* and $\hat{J}$) were able to correlate the time-dependent tests. The three integrals ($\Delta T_p{}^*$, J^*, and $\hat{J}$) were able to correlate the data for Hastelloy-X from the first contract under isothermal and OP TMF loading reasonably well. Finally, the buttonhead single edge crack specimen proved an excellent choice for most of the second program, with the possible exception of the temperature gradient tests.

REFERENCES

Ainsworth, R.A., Neale, B.K., and Price, R.H., 1978, "Fracture Behavior in the Presence of Thermal Strains," *Proc. Inst. of Mech. Engrs., Conf. on Tolerance of Flaws in Pressurized Components*, London, pp. 171-178.

Atluri, S.N., Nishioka, T., and Nakagaki, M., 1983, "Incremental Path-Independent Integrals in Inelastic and Dynamic Fracture Mechanics," Report GIT-CACM-SNA-83-27, Georgia Inst. of Tech., Atlanta GA.

Blackburn, W.S., 1972, "Path-Independent Integrals to Predict Onset of Crack Instability in an Elastic Material," *Int. J. Fracture*, Vol. 8, pp. 343-346.

Brust, F.W., Nakagaki, M., and Springfield, C., 1989, "Integral Parameters for Thermal Fracture," *Eng. Fract. Mech.*, Vol. 33, No. 4, pp. 561-579.

Dugdale, D.S., 1960, "Yielding of Steel Sheets Containing Slits," *J. Mech. Phys. Solids*, Vol. 8, pp. 100-104.

Elber, W., 1971, "The Significance of Fatigue Crack Closure," *Damage Tolerance in Aircraft Structures*, STP 486, ASTM, pp. 230-242.

Erdogan, F., and Ratwani, M., 1970, "Fatigue and Fracture of Cylindrical Shells Containing a Circumferential Crack," *Int. J. Fracture*, Vol. 6, No. 4, pp. 379-392.

Gurtin, M.E., "On a Path-Independent Integral for Thermoelasticity," 1979, *Int. J. Fracture*, Vol. 15, pp. R169-R170.

Goldman, N.L., and Hutchinson, J.W., 1975, "Fully Plastic Crack Problems: The Center-Cracked Strip Under Plane Strain, *Int. Jour. Solids and Struct.* Vol. 11, pp. 575-591.

Iyyer, N.S., and Dowling, N.E., 1985, "Closure of Fatigue Cacks at High Strains,", NASA CR 175021.

Jordan, E.H., and Meyers, G.J., 1986, "Fracture Mechanics Applied to Nonisothermal Fatigue Crack Growth," *Eng. Fracture Mech.*, Vol. 23, No. 2, pp. 345-358.

Jordan, E.H., and Meyers, G.J., 1989, "Fracture Mechanics Applied to Elevated Temperature Crack Growth," *J. Engr. Matls. & Tech.*, Vol. 111, pp. 306-313.

Kim, K.S., 1985, "A Review of Path-Independent Integrals in Elastic-Plastic Fracture Mechanics," NASA CR-174956.

Kim, K.S., and Orange, T.W., 1988, "A Review of Path-Independent Integrals in Elastic-Plastic Fracture Mechanics," *Fracture Mechanics: Eighteenth Symposium*, STP-945, ASTM, pp. 713-729.

Kim, K.S., Van Stone, R.H., Malik, S.N., and Laflen, J.H., 1990a, "Elevated Temperature Crack Growth," NASA CR 182247.

Kim, K.S., Van Stone, R.H., Laflen, J.H., and Orange, T.W., 1990b, "Elevated Temperature Crack Growth Simulation with Closure," *Fracture Mechanics: Twenty-first Symposium*, STP-1074, ASTM, pp. 421-447.

Kim, K.S., and Van Stone, R.H., 1990, "Application of Path-Independent Integrals to Elevated Temperature Crack Growth," *Elevated Temperature Crack Growth*, Proc. Symposium, ASME Winter Ann. Mtg., MD-18, ASME, pp. 155-167.

Kim, K.S., and Van Stone, R.H., 1992, "Elevated Temperature Crack Growth," NASA CR 189191.

King, R.B., and Herrmann, G., 1981, "Nondestructive Evaluation of the J and M Integrals," 1981, *J. Appl. Mech.*, Vol. 48, March 1981, pp. 83-87.

Kishimoto, K., Aoki, S., and Sakata, M., 1980, "On the Path-Independent Integral - Ĵ", *Eng. Fract. Mech.*, Vol. 13, No. 4, pp. 841-850.

Landes, J.D., and Begley, J.A., 1976, "A Fracture Mechanics Approach to Creep Crack Growth," *Mechanics of Crack Growth*, STP-590, ASTM, pp. 128-148.

Malik, S.N., Van Stone, R.H., Kim, K.S., and Laflen, J.H., 1987, "Elevated Temperature Crack Growth - Annual Report," NASA CR-179601.

Meyers, G.J., 1982, "Fracture Mechanics Criteria for Turbine Engine Hot Section Components," NASA CR-167896.

Rice,J.R., 1968, "A Path-Independent Integral and the Approximate Analysis of Strain Concentration by Notches and Cracks," *J. Appl. Mech.*, Vol. 35, pp. 379-386.

Shih, C.F., and Hutchinson, J.W., 1976, "Fully Plastic Solutions and Large Scale Yielding Estimates for Plane Stress Crack Problems," *J. Engr. Matls. & Tech.*, Vol. 98, October 1976, pp. 289-295.

Tomkins, B., 1975, "The Development of Fatigue Crack Propagation Models for Engineering Applications at Elevated Temperatures," *J. Eng. Matls. & Tech.*, Vol. 97, October 1975, pp. 289-297.

Wilson, W.K., and Yu, I.W., 1979, "The Use of the J-Integral in Thermal Stress Crack Problems," *Int. J. Fracture*, Vol. 15, No. 4, pp. 377-387.

Yau, J.F., Malik, S.N., Kim, K.S., Van Stone, R. H., and Laflen, J.H., 1985, "Elevated Temperature Crack Growth - Annual Report," NASA CR-174957.

AD-Vol. 36, Fatigue and Fracture of Aerospace Structural Materials
ASME 1993

FATIGUE PROPERTIES OF COATED TWO-DIMENSIONAL CARBON-CARBON COMPOSITES

Ajit K. Roy
University of Dayton Research Institute
Dayton, Ohio

ABSTRACT

The room-temperature in-plane fatigue properties of coated two-dimensional $[0/90]_{2S}$ carbon-carbon composites are reported. The properties are obtained in tension-tension, compression-compression, and tension-compression fatigue loading. It is observed that the fatigue property degradation is mostly due to matrix damage. The extent of the matrix damage is determined by measuring the interlaminar tensile strength of the fatigue specimens. The influence of the above fatigue loading on the interlaminar tensile strength of the composite is reported.

INTRODUCTION

For structural application of two-dimensional carbon-carbon composites, the performance of the material under fatigue loading needs to be characterized in order to assess the durability of the material under sustained loading. At present, very limited fatigue data of two-dimensional carbon-carbon composites are available in the literature, for example, Ozturk and Moore[1]. They compared the tensile fatigue behavior of two plain weave carbon-carbon composites produced under two different processes. In this work the in-plane static and low-cycle (10 Hz) fatigue properties of two-dimensional carbon-carbon composite laminates are presented. The composite was made of 8HS balanced fabric of 3K tow of Amoco T300 fiber with 24 tows per inch in the warp direction and 23 in the fill direction. The composite contained an inhibited matrix, and during processing the composite was heat treated and the matrix was multicycle CVI densified. Rectangular specimen blanks of size 204x25.4 mm aligned in the warp direction were cut from the $[0/90]_{2S}$ laminated panel. Each test specimen blank was coated with slurry sealout and then the CVD SiC coating was applied outside.

TEST SPECIMENS

The test specimens were prepared from the rectangular blanks by configuring to a dog-bone shape as shown in Figure 1. Since the matrix of the composite was inhibited, the specimens were cut dry using a diamond saw and then machined with a diamond impregnated grinder. The width of the specimen in the gage section was reduced to 50 percent of the total width of the blanks for direct gripping of the specimens at the ends without using any end tabs. For a $[0/90]_{2s}$ laminate of 8HS fabric with 24 tows in the warp and 23 tows in the fill direction, the size of a unit cell size of the fabric is approximately 8.8 by 8.8 mm. Thus the gage width of the specimens was larger than a unit cell size of the fabric. The gage length of the specimens contained more than seven unit cells of the fabric.

FATIGUE TESTS

The objective of this work was to obtain the low-cycle fatigue behavior of the composite and assess the fatigue damage. Load-controlled fatigue tests were performed in tension-tension, compression-compression, and in tension-compression, in air and at room temperature using a servo-hydraulic MTS testing machine. All the fatigue tests were performed with sinusoidal waveform at a frequency of 10 Hz. The stress ratio, R (the ratio between the minimum stress to the maximum stress in the wave form), of tension-tension fatigue was 0.1, for compression-compression fatigue was 10, and that for tension-compression fatigue was -1.0. All the fatigue tests were run up to 10^6 cycles or until specimen failure if the specimen failure occurred before 10^6 cycles. The tension-tension fatigue tests were performed at four different stress levels. Due to limited availability of the specimens, the compression-compression and tension-compression tests were run at one stress level. To determine the stress level of the fatigue tests, the static stress-strain curves and the strengths in tension and compression were obtained. The scanning electron microscopy (SEM) of the fractured surface was performed for microstructural analysis of the fatigue damage.

RESULTS AND DISCUSSIONS

The representative static tensile stress-strain curves of the composite are shown in Figure 2. The two values of the tensile stiffness shown in the figure are the initial stiffnesses of the two specimens tested. There was a scatter of about eight percent on the measured initial tensile stiffness of the five specimens tested which indicates some material variability of the laminate. The average tensile strength of the composite was 235 MPa (34 ksi). The tensile failure strain varied by a great deal from specimen to specimen as shown in Figure 2. There exists a knee in the stress-strain curves of the specimens at about 85 percent of the failure strength which indicates substantial damage in the material (may be attributed to failure some of the fiber yarns in the vicinity of the high crimp angle) at that stress level.

Based on the above observation on the presence of knee in the stress-strain curve, the tension-tension fatigue was performed at three stress levels: 70, 80, 85, and 90 percent of the tensile strength. The residual tensile stress-strain curves of two tension-tension fatigue run-out specimens at 10^6 cycles, and at stress level, S, of 70 percent of ultimate tensile strength, are compared with that of the static tensile test in Figure 3. Although there was stiffness degradation due to fatigue, the knee in the stress-strain curve of the fatigue run-out specimens practically remained at the same value of the stress as that of the static specimens. This implied that at the stress level of 70 percent of the ultimate strength in tension-tension fatigue, the fatigue damage was mainly limited to the matrix damage. In addition, to monitor the stiffness degradation due to fatigue damage with increasing number of cycles, the stress-strain curves of the fatigue specimens at 10^2, 10^4, and 10^6 cycles were obtained. The stiffness change due to fatigue loading at S=0.7 with increasing fatigue cycles was found to be very small, as shown in Figure 4. To assess the tension-tension fatigue damage at different stress levels, the stiffness change with fatigue cycles of the specimens at three different stress levels is shown in Figure 5. It was obvious from the figure that the stiffness change due to the tension-tension fatigue was limited to 20 percent, even at a stress level of 90 percent of the fatigue run-out specimens. If the stress level was limited to 80 percent of the ultimate load, the stiffness reduction was less than 15 percent of the initial stiffness. The residual strength of the fatigue run-out specimens was found to be about 98 percent of the static strength. Figure 6 shows the S-N curve of the tension-tension fatigue. From the S-N curve it is also observed that at and below a stress level of 80 percent of the tensile strength, the fatigue life of all the test specimens was more than 10^6 cycles. Thus it appears that the carbon-carbon laminate has an excellent fatigue life in tension-tension fatigue even at a stress level of 80 percent of static tensile strength.

The static stress-strain curves, both in tension and compression, are shown in Figure 7. The initial stiffness in compression is about two percent lower than that in tension. The lower stiffness in tension may be due the fiber waviness present in the fabric. The fatigue test in compression was performed with specimen side supported to prevent the specimens from buckling. The static compressive strength of the laminate was found to be 18 ksi which is about 53 percent of that in tension. All the tests in compression-compression fatigue, as mentioned before, were carried out at a stress level of 80 percent of the static compressive strength (S=0.8). The stress-strain curves at several fatigue cycles of a representative compression-compression fatigue specimen are shown in Figure 8, and the stiffness change with increasing number of fatigue cycles is shown in Figure 9. Figure 9 revealed that although for some specimens the stiffness decreased with increasing fatigue cycles, the stiffness for two other specimens increased with increasing fatigue cycles. The phenomenon of the stiffness increasing with fatigue cycles, however, was not observed in tension-tension fatigue. All the specimens in compression-compression fatigue (at S=0.8) had run out 10^6 cycles. Consequently, since all the tests in compression-compression fatigue were carried out at one stress level of S=80, adequate data was not generated to obtain the relevant S-N curve. Further, the residual compressive strength of the fatigue run-out specimens in C-C fatigue was found to be about 95 percent of the compressive strength.

In addition to the T-T and C-C fatigue, fully reversal ($\sigma_{max} = -\sigma_{min}$; R=-1.0) tension-compression (T-C) fatigue testing was carried out. Because the compressive strength of the laminate was lower than that in tension, the stress level of the test, S, was set at 80 percent of the compressive strength. The stress-strain curves of a representative specimen before fatigue and after T-C fatigue run out is shown in Figure 10. In this case the tensile stiffness reduction was about five percent of the tensile stiffness, and the compressive stiffness reduction was less than two percent of the compressive stiffness. The T-C fatigue test was performed only at S=0.8, and all the specimens survived up to 10^6 cycles. Thus with the limited data, the S-N curve of the T-C fatigue is not shown.

The fatigue stiffness reduction data of the fatigue run-out specimens indicated that the fatigue damage may be mostly limited to the matrix damage. To assess the matrix damage, the interlaminar tension (ILT) test [2] was carried out on the buttons specimens prepared from the gage section of the fatigue test specimens, as shown in Figure 11. Further, to investigate the cause of the stiffness increase with fatigue cycles in C-C fatigue, the C-C fatigue ILT specimens were prepared from the specimens whose stiffness increased with increasing fatigue cycles (see Figure 9). The ILT strengths of the fatigue specimens were compared with that of the virgin specimens. Since all the test specimens were coated, the ILT specimens with coating had the carbon-carbon substrate sandwiched in between two surface coatings. All the ILT specimens with coating failed at the interface between the coating and the carbon-carbon substrate. The ILT strength data in Figure 12 are the failure strength of the coating interface. The interface ILT strength of the T-T and T-C fatigue specimens was about 40 percent lower than that of the virgin specimens, whereas that of the C-C fatigue specimens (whose stiffness increased with the fatigue cycles) appeared to be almost the same as the virgin specimens. Thus the tension fatigue seemed to cause more coating interface damage than the compression.

To assess the fatigue matrix damage, the coating on the ILT specimens used above for Figure 12 was removed and the ILT test was repeated. One of the surface coatings of the coated ILT specimens failed in obtaining data shown in Figure 12. The other surface coating was sliced out by gently tapping a razor blade through the interface between the coating and the carbon-carbon substrate. The ILT strengths of the fatigue run-out carbon-carbon laminate are shown in Figure 13. The strength data in Figure 13 indicated there was a degradation of ILT strength of the carbon-carbon laminate (which is mostly attributed to the matrix damage) due to fatigue. The matrix damage due to T-T fatigue seemed to be more severe than the other two fatigue loadings.

CONCLUSIONS

It has been observed from the above fatigue stiffness reduction data and fatigue residual strength data that the carbon-carbon $[0/90]_{2s}$ laminate tested has very good fatigue properties. In the case of the T-T fatigue test, at a stress level of 80 percent of the tensile strength, the fatigue run-out specimens retained about 98 percent of the static tensile strength, and the fatigue stiffness reduction was less than 15 percent of the static stiffness. In C-C fatigue some specimens showed increased stiffness with increasing fatigue cycles. The residual strength of the C-C fatigue run-out specimens was about 95 percent of the static compressive strength. An ILT test was used to assess the fatigue matrix damage.

ACKNOWLEDGMENT

This work was sponsored by the WL Materials Directorate under contract number F33615-91-C-5618. The financial support is gratefully acknowledged.

REFERENCES

1. Ozturk, A and R.E. Moore, "Tensile Fatigue Behaviour of Tightly Woven Carbon/Carbon Composites," Composites, Vol 23, No 1, Jan 1992, pp 39-46.

2. Roy, A.K., "A Self-Aligned Test Fixture for Interlaminar Tensile Testing of Two-Dimensional Carbon-Carbon Composites," presented at the 17th Mechanics of Composites Review, Dayton, OH, October 27-28, 1992.

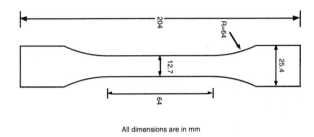

All dimensions are in mm

Figure 1. Configuration and dimensions of the test specimens.

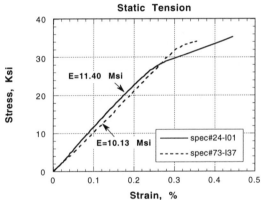

Figure 2. Stress-strain curves of [0/90]$_{2s}$ carbon-carbon laminate in static tension.

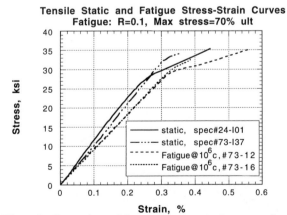

Figure 3. Comparison of the tensile stress-strain curves of static specimens with that of the tension-tension fatigue specimens run out after 10^6 cycles at stress level of 70 percent of static strength.

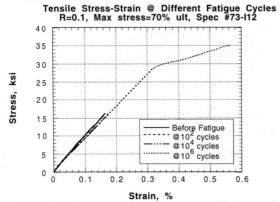

Figure 4. Tensile stress-strain curves at several fatigue cycles of a tension-tension fatigue specimen at stress level of 70 percent of tensile strength.

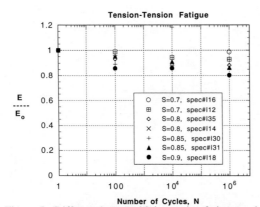

Figure 5. Stiffness change with increasing fatigue cycles of tension-tension fatigue.

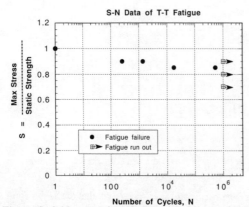

Figure 6. S-N curve of tension-tension fatigue of [0/90]$_{2s}$ carbon-carbon laminate.

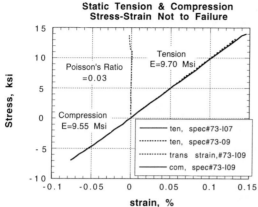

Figure 7. Static stress-strain curves in tension and compression of [0/90]$_{2s}$ carbon-carbon specimens.

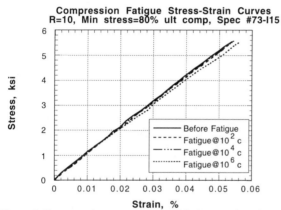

Figure 8. Stress-strain curves at several fatigue cycles of a compression-compression fatigue specimen at stress level of 80 percent of compressive strength.

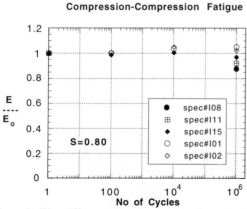

Figure 9. The stiffness versus the compression-compression fatigue at stress level of S=0.8.

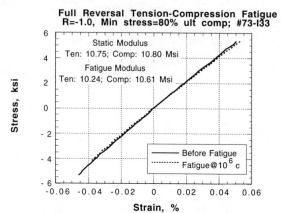

Figure 10. Stress-strain curves of fully reversal tension-compression fatigue (R=-1.0).

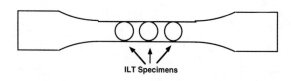

Figure 11. Location of the interlaminar tensile (ILT) button specimens obtained from the fatigue specimens for assessing fatigue damage.

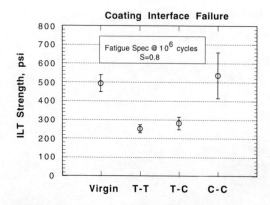

Figure 12. The ILT strengths of coating and carbon-carbon interface of the button specimens obtained from virgin and fatigue specimens. The data is the average of six test specimens in each category.

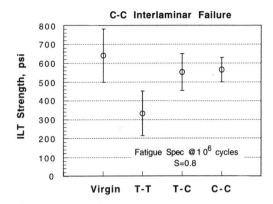

Figure 13. The ILT strengths of carbon-carbon substrate of the button specimens obtained from virgin and fatigue specimens. The data is the average of six test specimens in each category.

AD-Vol. 36, Fatigue and Fracture of Aerospace Structural Materials
ASME 1993

FAILURE MODES IN CUBIC CRYSTALLINE MATERIALS

M. A. Zikry
Department of Mechanical and Aerospace Engineering
North Carolina State University
Raleigh, North Carolina

ABSTRACT

The micromechanical failure mechanisms of high-strain-rate void collapse in f.c.c. and b.c.c. single crystals were studied. A constitutive formulation was developed based on the micromechanisms of crystalline plasticity. An explicit dynamic finite-element computational algorithm was formulated for the integration of the numerically stiff micromechanical constitutive relations. Criteria for the formation and the exclusion of shear bands are presented. An increase in the strain hardening, the rate sensitivity, and the strain-rate hardening of the crystal resulted in the preclusion of shear band formation. Material failure, in this case, is due to the large stress concentrations, high temperatures, and large plastic slip strain-rates near the tip of the collapsed void. Mechanisms for the formation of tensile cracks were studied; and comparisons were made with experimental observations. The resolved shear stresses, the temperature, and the plastic slip strain-rates, are shown to be physically limited by the dislocation densities and velocities of each cubic lattice structure.

1. INTRODUCTION

Microscopic voids play a crucial role in the failure of brittle and ductile materials. Ductile fracture may occur when the voids grow and coalesce. In other cases, localization of the plastic deformations occurs before general coalescence takes place. Failure by void growth, where the material is subjected to tensile loads, has been studied more extensively than the corresponding problem of void collapse, where the material is subjected to compressive loads. Carroll and Holt (1971) conducted a theoretical and experimental study on the development of static and dynamic pore-collapse relations for ductile porous materials. They obtained static and dynamic pore-collapse relations for the compression-induced collapse of a hollow sphere in an incompressible elastic-plastic material subjected to shock-loading conditions. Butcher (1973) studied the dynamic-shock response of partially compacted porous aluminum during the unloading of the specimen. Butcher et al. (1974) have developed criteria for the influence of pore size, inertia, and viscosity on the compaction of aluminum. Johnson (1981) introduced a mathematical model for ductile hole growth under the application of dynamic tensile stresses.

Essentially all the above referenced analytical models, for void growth and void collapse, have been based on phenomenological constitutive relations. Hori and Nemat-Nasser (1988) presented analytical micromechanical plane-strain and three-dimensional models for void growth and void collapse for a rate-dependent material subjected to high rates of strain. Hori and Nemat-Nasser (1988) accounted for the local anisotropic plastic flow by using a micromechanically based single crystal plasticity model. Nemat-Nasser and Chang (1990) performed a series of experiments to study the compression-induced dynamic collapse of monocrystalline and polycrystalline copper. They showed that *tensile* cracks can form normal to the direction of the compression stress-axis. After a period of incubation, the cracks dynamically propagate and this eventually results in the failure of the monocrystalline and polycrystalline copper structures. Recently, Zikry (1993a) developed a model to investigate the phenomena of tensile crack formation in f.c.c. crystals observed by Nemat-Nasser and Chang (1990). The major objective of the present study is to extend the physically based analysis, developed by Zikry (1993a), to gain a detailed understanding of the material failure modes associated with the high strain-rate deformation of cubic metallic crystals, specifically, f.c.c. and b.c.c. single crystals.

2. FORMULATION

In this section, the constitutive relations for the high strain-rate finite deformation of rate-dependent single crystals are introduced.

The velocity gradient, $V_{i,j}$, is decomposed into its symmetric and anti-symmetric parts,

$$V_{i,j} = D_{ij} + W_{ij}, \tag{1}$$

where D_{ij} is the deformation rate tensor,

$$D_{ij} = 1/2 \, (V_{i,j} + V_{j,i}), \tag{2}$$

and W_{ij} is the spin tensor,

$$W_{ij} = 1/2 \, (V_{i,j} - V_{j,i}). \tag{3}$$

The total deformation tensor, D_{ij}, and the total spin tensor, W_{ij}, are each decomposed into an elastic part and a plastic part,

$$D_{ij} = D_{ij}^* + D_{ij}^p, \tag{4}$$

$$W_{ij} = W_{ij}^* + W_{ij}^p. \tag{5}$$

The superscript * denotes the elastic part and the superscript p denotes the plastic part; W_{ij}^* includes the rigid body spin. The inelastic parts are defined in terms of the crystallographic slip-rates as

$$D_{ij}^p = P_{ij}^{(\alpha)} \, \dot{\gamma}^{(\alpha)}, \tag{6}$$

$$W_{ij}^p = \omega_{ij}^{(\alpha)} \, \dot{\gamma}^{(\alpha)}, \tag{7}$$

where α is summed over all slip systems, and the tensors $P_{ij}^{(a)}$ and $\omega_{ij}^{(a)}$ are given by

$$P_{ij}^{(\alpha)} = 1/2 \, (s_i^{(\alpha)} n_j^{(\alpha)} + s_j^{(\alpha)} n_i^{(\alpha)}), \tag{8}$$

$$\omega_{ij}^{(\alpha)} = 1/2 \, (s_i^{(\alpha)} n_j^{(\alpha)} - s_j^{(\alpha)} n_i^{(\alpha)}), \tag{9}$$

where $n_i^{(\alpha)}$ is the unit vector normal to the slip plane, $s_i^{(\alpha)}$ is the unit vector in the slip direction and there is no sum on α. For a rate-dependent formulation, the slip-rates are functions of the resolved shear stresses and the reference stress. The expression used here is of the power law form,

$$\overset{.}{\gamma}^{(\alpha)} = \overset{.}{\gamma}_r^{(\alpha)} \left[\frac{\tau^{(\alpha)}}{\tau_r^{(\alpha)}} \right] \left[\left| \frac{\tau^{(\alpha)}}{\tau_r^{(\alpha)}} \right| \right]^{\frac{1}{m} - 1} \qquad \text{(no sum on } \alpha \text{)} ,\qquad (10)$$

where $\overset{.}{\gamma}_r^{(\alpha)}$ is a reference shear-strain-rate corresponding to a reference shear stress $\tau_r^{(\alpha)}$. The reference stress for a crystalline material is given by

$$\tau_r^{(\alpha)} = F(\gamma) \left[\frac{T}{T_r} \right]^{-\nu} , \qquad (11)$$

where $F(\gamma)$ is a strain-hardening function, and it is generally formulated to fit the specific dynamic strain-hardening response of the material. T is the current temperature of the solid, T_r is the reference temperature, ν is a thermal softening parameter which is material dependent, $\tau^{(\alpha)}$ is the resolved shear-stress on the αth slip-system and the exponent m, is the material rate sensitivity parameter. The rate sensitivity parameter, m, is material-dependent and has important implications in high strain-rate analyses. For values of the shear slip-rate smaller than a critical value; the lattice motion is thermally activated, (Follansbee et al., 1984). The rate sensitivity parameter is approximately equal to one for slip-rates greater than the critical slip-rate. In this case, the flow is characterized by drag-controlled dislocation motion.

From the balance of energy, with no thermal conduction, the time rate of change in temperature is related to the rate of the plastic work by

$$\overset{.}{T} = \frac{\chi}{\rho c_p} D_{ij}^P \sigma_{ij}' , \qquad (12)$$

where χ is the fraction of the plastic work converted to heat, σ_{ij}' is the deviatoric stress, ρ is the material density, and c_p is the specific heat of the material. Due to the high strain-rate deformation of the crystal, the rate of change of temperature is strictly a function of adiabatic heating This can be qualitatively understood from the Fourier modulus, (Culver, 1973),

$$\Gamma = \frac{k}{\rho \, c_p L_0^2 \, \overset{.}{\varepsilon}} , \qquad (13)$$

where k is the thermal conductivity, c_p is the heat capacitance, ρ is the density, and L_0 is the initial length of the specimen.

3. NUMERICAL METHOD

The total deformation tensor, D_{ij}, and the plastic deformation tensor, D_{ij}^P, are needed to update the stress state of the crystalline material. A brief outline of the numerical method will be presented, for further details see Zikry (1993b). In stress wave propagation problems, the resolution of the high frequency response of wave reflection and diffraction requires time-steps of the order of the time it takes the wave to transverse an element. This is of the same order, as the stability limit for explicit schemes, and hence, explicit finite-element methods are preferred for high frequency response problems. The equations of motion are integrated by the explicit central difference method. Four-node quadrilateral elements are used with one-point Gauss integration. Once the nodal accelerations, velocities, and displacements are obtained, the total deformation rate tensor, D_{ij}, and the total spin tensor, W_{ij}, can be calculated at each time increment.

To solve for the plastic deformation rate tensor, D_{ij}^P, the time derivative of the resolved shear-stress,

$$\overset{.}{\tau}^{(\alpha)} = \frac{d}{dt} \, (P_{ij}^{(\alpha)} \sigma_{ij}) , \qquad (14)$$

is used together with the Jaumann stress rate, an objective stress rate (Zikry and Nemat-Nasser, 1990) and the assumption that the elastic modulus tensor is isotropic, to obtain the following system of nonlinear differential equations,

$$\dot{\tau}^{(1)} = 2\mu \left(P_{ij}^{(1)} D_{ij} - P_{ij}^{(1)} \left[P_{ij}^{(1)} \dot{\gamma}_r^{(1)} \left(\frac{\tau^{(1)}}{\tau_r} \right)^{\frac{1}{m}} + P_{ij}^{(2)} \dot{\gamma}_r^{(2)} \left(\frac{\tau^{(2)}}{\tau_r} \right)^{\frac{1}{m}} \right] \right), \tag{15}$$

$$\dot{\tau}^{(2)} = 2\mu \left(P_{ij}^{(2)} D_{ij} - P_{ij}^{(2)} \left[P_{ij}^{(1)} \dot{\gamma}_r^{(1)} \left(\frac{\tau^{(1)}}{\tau_r} \right)^{\frac{1}{m}} + P_{ij}^{(2)} \dot{\gamma}_r^{(2)} \left(\frac{\tau^{(2)}}{\tau_r} \right)^{\frac{1}{m}} \right] \right), \tag{16}$$

for a double slip planar model. It is also assumed that the lattice spin is given by

$$\dot{n}^{(\alpha)} = W_{ij}^* n_j, \qquad\qquad \dot{s}_i^{(\alpha)} = W_{ij}^* s_j^{(\alpha)} \tag{17}$$

$$\psi(t + \Delta t) = \psi(t) + \dot{\psi} \Delta t, \tag{18}$$

and the time derivative of the angle is measured by the lattice spin

$$W_{ij}^* = \dot{\psi} \begin{bmatrix} 0 & -1 \\ 1 & 0 \end{bmatrix}. \tag{19}$$

The solution to the system of ordinary differential equations, (15) and (16), is numerically difficult, not only due to the nonlinearity of the resolved shear stress, but also because the system of equations is numerically stiff in certain time intervals. The numerical stiffness is caused by different time scales of the resolved shear stress. This results in eigenvalues corresponding to the Jacobian of the initial value problem, that are widely varying. This leads to the growth of numerically propagated error, i.e. instability in the solution of the system of differential equations. To deal with this problem, an algorithm is proposed. A brief outline of the algorithm is given below.

Since the pair of equations, (15) and (16), is only stiff in some regions of the integration domain, an explicit fifth-order accurate Runge-Kutta method is used over most of the time domain. The propagated error is measured by the growth in the local truncation error. If the time-step must be restricted due to stability, and not accuracy, a backward Euler method is used. The backward Euler method is both A-stable and stiffly stable; it is also an order-one Backward Differentiation Formula. The trapezoidal rule, which is also A-stable and second-order accurate, has integration domains where the numerical solutions may become unstable for certain non-negative real eigenvalues (Gear, 1971). The algorithm methodology is as follows:

Automatic step control is achieved by using step doubling on the Runge-Kutta fourth-order method. Two approximate solutions are taken, one solution of step size 2h and a second solution with two steps, each of size h.

$$\tau(t+2h) = \hat{\tau}_1 + (2h)^5 \phi + O(h^6) + \dots, \tag{20}$$

$$\tau(t+2h) = \hat{\tau}_2 + 2(h)^5 \phi + O(h^6) + \dots, \tag{21}$$

Since the original method is fourth-order, the two numerical methods are combined give a solution of fifth-order accuracy

$$\tau(t+2h) = \hat{\tau}_2 + \frac{\Delta}{15} + O(h^6), \tag{22}$$

where $\Delta = \hat{\tau}_1 - \hat{\tau}_2$ is the local truncation error which measures how well the solution is approximated at each time-step. Based on this error measure, an adjusted time-step is calculated,

$$h_{new} = F\, h_{old} \left| \frac{\Delta_0}{\Delta_1} \right|^{0.20}, \tag{23}$$

where h_{new} is the adjusted time-step, and h_{old} is the initial time-step. Δ_1 is the actual accuracy measured by the supremum norm as max $|\tau_1 - \tau_2|$, and Δ_0 is the desired accuracy measured by εH. Here ε is the tolerance level supplied by the user and H is a scaling factor for fractional errors for the ith equation given by $|\tau| + |h\frac{d\tau}{dt}|$, where h is the initial time-step. The factor F serves to keep the new time-step small enough to be accepted, if the truncation error in the next time-step is growing. Based on (23), the time-step is increased if the truncation error is smaller than the desired accuracy; and conversely the time-step is decreased if the truncation error is greater than the desired accuracy.

Since Runge-Kutta methods have finite stability regions, there can be a growth in the propagated error, and, therefore, the time-step, in certain time domains, is restricted due to stability and not the accuracy requirement given by εH. This is an indication of stiff behavior. In the present algorithm, the largest allowable time-step is chosen, .i.e., the time-step on the stability boundary. This implies that the local errors are of the same magnitude as the accuracy tolerance used in (23). If the time-step is unduly restricted due to stability, the solution will proceed in time, albeit inefficiently, due to the necessity of using intolerably small time-steps. To correctly identify the regions of numerical stiffness, and to distinguish a step reduction due to *accuracy* from a time-step reduction due to *stability*, a stiffness ratio, S_R, is proposed. It is defined as

$$S_R = \frac{|Re \, \lambda|_{max}}{|Re \, \lambda|_{min}} \left(\frac{1}{t_2 - t_1}\right), \tag{24}$$

where $|Re \, \lambda|_{max}$ and $|Re \, \lambda|_{min}$ are the greatest and smallest absolute values of the real parts of the eigenvalues of the Jacobian of the system of ordinary differential equations, and $t_2 - t_1$ is the time interval of the integration. A large stiffness ratio, S_R, signifies that the ratios of the eigenvalues are dispersed relative to the time scale. When the time-step is restricted due to the presence of the widely varying eigenvalues, this is a stability problem and an indication that the initial-value problem is numerically stiff. The eigenvalues of the Jacobian of the system of equations, (29) and (30), are given by

$$\lambda_{max,min} = - B \pm (B^2 \pm 4 A C)^{1/2} \tag{25}$$

where A is $\dot{\gamma}^{(1)} \dot{\gamma}^{(2)}$, B is $\dot{\gamma}^{(1)} + \dot{\gamma}^{(2)}$, C is $\sin^2(4\phi)$, and ϕ is the initial crystal orientation. Since the eigenvalues are functions of the shear strain-rates, $\dot{\gamma}$, and the initial crystal orientation, ϕ, the stiffness ratio, S_R, can also be used a localization parameter to characterize the onset of the shear-strain localization of the single crystal. An increasing stiffness ratio is an indication that for a specified deformation mode, the slip-rates, $\dot{\gamma}$, are much greater for one slip system than for the other active slip systems. Namely, one of the slip systems is dominating the deformation process. The domination of one slip system, over other active slip systems, usually occurs, when a macroscopic shear band is forming in the solid. As the analyses will show, shear bands form at orientations that are almost coincident with the orientation of one dominant slip system.

If the time-step becomes much smaller than the Courant number and the stiffness ratio is increasing, then this is an indication of numerical stiffness, since the time-step is being reduced due to stability considerations. In the present analysis, the integration is automatically switched to the backward Euler scheme, for increasing values of the stiffness ratio and decreasing values of the time-step. The quasi-Newton method is used to solve the system of nonlinear algebraic equations.

4. RESULTS

The constitutive model, for the high strain-rate deformation of the single crystal, and the dynamic explicit finite-element algorithm were applied to study the collapse of an initially circular single void, in a square plate, subjected to compressive nominal strain-rates. The geometrical dimensions were chosen to approximately match the specimen dimensions used by Nemat-Nasser and Chang (1990) in their experimental study of the compression-induced collapse of a void in a single copper crystal. Hence, the half-length of the plate, L, was chosen as 5 mm and the radius of the circular void, R, was chosen as one-tenth of the length, Fig. 1a. For an initially symmetric double mode of slip, the resultant double slip system for the active four slip systems, reported in the experiment by Nemat-Nasser and Chang (1990), are the $[\bar{2}1\bar{1}]$ $(\bar{1}11)$ system, which was chosen as the primary slip system, and the $[\bar{2}11]$ $(1\bar{1}1)$ system which was chosen as the conjugate slip system, Fig. 1b. The material properties were chosen to be representative of single f.c.c. copper crystals. Therefore, Young's modulus was taken as 1000 τ_y, where τ_y is the static yield, and it was chosen as 110 MPa. In this study, the following material properties were used: 0.30 for Poisson's ratio, 8900 kg/m^3 for the mass density, .02 for the strain-rate sensitivity, and a reference slip strain-rate of 10^{-3}/s for both slip systems. The critical slip strain-rate was chosen as 10^4/s, which is representative of copper (Follansbee et al., 1984).

To account for this high rate of dynamic strain hardening, the following reference stress, τ_r, introduced by Zikry (1993a) for the αth slip system, is proposed:

$$\tau_r^{(\alpha)} = \tau_y \left(1 + D \gamma^n \right) \left[\frac{T}{T_r} \right]^{-\nu} , \qquad (26)$$

where γ is the accumulated plastic slip strain on all the slip systems, D is a constant hardening parameter, n is a strain hardening exponent, ν is the thermal softening exponent, and T_r is the reference temperature. The reference temperature was chosen as 20^0 C for all calculations in this analysis. To study the effects of strain hardening and thermal softening on the collapse of the void, and to match the experimental results of Nemat-Nasser and Chang, the hardening and softening parameters for the reference stress were varied as follows:

Table 1
Parameters for the Reference Stress for the Copper Single Crystal

D	n	ν
8.0	.10	.50
24.0	.10	.40
32.0	.10	.40
42.0	.10	.40
48.0	.10	.30
64.0	.15	.30
100.0	.20	.30

The values chosen for ν and n are typical for f.c.c. crystals (Klopp et al., 1985). Taylor hardening was used in the present study; i.e., it was assumed that both slip systems harden at the same rate. The finite-element mesh was designed such that the aspect ratio of each element of the square plate is initially square (Fig. 2a). Based on a convergence analysis for nominal strain-rates ranging from 200/s to 10^4/s, 1872 four-noded quadrilateral elements were used in the present analysis for a nominal strain-rate of 1000/s. The average stress-strain curves, for different values of the reference stress, at a nominal compressive strain-rate of 1000/s, are shown in Fig. 1c. The load carrying capacity of the crystal increased with increasing values of the reference stress (Fig. 2c). The load per unit undeformed area, σ_{avg}, is given by

$$\sigma_{avg} = \frac{1}{A} \int_0^A \sigma_{2i} \, n_i \, dA, \qquad (27)$$

where A is the initial undeformed area, and $\sigma_{2i} n_i$ (sum on i) is the surface traction on the X_2 -face of the element. The average stress for the highest hardening case, the D100 model, was approximately nine times higher than the lowest hardening case, the D8 model. A transition region occurred at the curve for the D48 model. This region was approximately where the nominal average load ceased to globally unload. At the lower values of the reference stress, there was an unloading of the nominal loads. At the higher values of the reference stress, the global nominal load continued to increase, as the compressive end displacements increased. The increase in the global flow stress precludes the unloading of the single crystal, and this subsequently will not lead to the localization of the specimen (Zikry, 1993a). The stress-strain curve for the D48 model approximately matched the experimental observations reported by Nemat-Nasser and Chang (1990). The global stress saturated at approximately the same nominal strain of 9%, (Fig. 1c), reported by Nemat-Nasser and Chang (1990). To illustrate the effects of strain-hardening and thermal softening on the evolution of the different failure modes associated with the compression-induced collapse of the void in the single crystal, the D8 model and the D48 model were chosen for detailed analyses

At a nominal strain of 5%, for the D8 model, the mesh had started to bulge at the free boundary, but no localization patterns were discernible (Fig. 2b). At a nominal strain of 8%, the mesh had localized at an orientation of approximately 45 degrees from the compression stress axis (Fig. 2c). The shear band extended to the free boundary of the mesh, and at the plane of the free boundary intersecting with the shear band, the mesh had geometrically bulged. The void had also deformed from an initially circular shape into an ellipse with the major axis in the X_1 direction.

A more quantitative picture of the localization is given by the contours of the total plastic strains, the primary and the conjugate slip-rates, the temperature distribution, and the rotation of the lattice system. At a nominal strain of 5%, the maximum accumulated plastic strain was .45, and the maximum strains had not fully extended to the free boundary (Fig. 3a). At a nominal strain of 8%, the maximum plastic strains were at the center of the band, and attained a value of 1.00 (Fig. 3b). The lowest strain values were on the outside of the band, and had a value of .25. Outside of the localized region, the mesh was strain-free. If the region outside the band were not strain-free, the band would not form, since there would be no discontinuities in the strain field. The primary slip-rates, at a nominal strain of 5%, are shown in Fig. 3c. The primary slip-rates were concentrated in a region oriented at approximately 45 degrees from the loading axis, and had a maximum value of 21.2×10^3/s (all slip-rates given in the contours were non-dimensionalized by the specimen length and the longitudinal elastic wave speed). At a nominal strain of 8%, the primary slip strain-rate attained a maximum of 26.9×10^3/s in the center of the band (Fig. 3d). At 5% nominal strain, the secondary slip-rates had a maximum value of 14.7×10^3/s (Fig. 3e). The secondary slip-rates, at a nominal strain of 8%, (Fig. 3f) had a maximum of 18.6×10^3/s in the localized region. The maximum values for slip-rates occurred at the center of the band and at the tip of the void at the $X_1 = 0$ face. As these results indicate, at this stage of the deformation, the primary slip system dominated the slip process. Furthermore, both primary and secondary slip-rates had exceeded the critical slip-rate, and hence, the lattice flow was characterized by drag control dislocation motion at this stage of the deformation. This accumulation of plastic and plastic strain-rates is due to the geometrical and thermal softening of the crystalline structure. At a nominal strain of 8%, the lattice in the band had rotated 10 degrees towards the compression stress axis; i.e., the primary slip system had geometrically softened and rotated to the maximum shear stress axis of 45 degrees (Fig. 3h). This 45 degree orientation resulted in the resolved shear stresses achieving maximum values, and this correspondingly caused an increase in the plastic strains. Thermal softening of the crystal coupled with this geometrical softening accelerates the shear-strain localization of the material. A maximum temperature of 490^0 C, at 8% nominal strain, occurred near the free surface of the void (Fig. 3i) at an orientation of approximately 45 degrees (the maximum resolved shear-stress orientation) from the compression stress-axis. This temperature is approximately .45 of the material's melting point (Mecking and Gottstein, 1982).

The contours corresponding to the D48 model, the reference stress that approximately matched the experimental stress-strain response reported by Nemat-Nasser and Chang (1990), at a nominal strain of 9%, are shown in Figs. 4(a-e). As the plastic strain contours show, the increase in hardening had resulted in a decrease in the plastic strains as compared with the plastic strains of the D8 hardening model. The maximum value of the accumulated plastic strains was .46 (Fig. 4a). The plastic strains were concentrated in a region bounded by the free surface of the void and the $X_2 = 0$ boundary. The strain field did not extend to the free surface of the mesh, and the free surface of the mesh had not bulged. The primary slip strain-rate attained a maximum of 64.4×10^3/s at the tip of the collapsed void (Fig. 4b). The secondary slip-rate had a maximum value of 10.6×10^3/s (Fig. 4c). At this stage of deformation, the slip process was dominated by the primary slip system. The primary slip-system had rotated 7.3 degrees towards the compression stress axis; i.e., this system had geometrically softened, and had rotated in the direction of the maximum shear stress axis of 45 degrees. This lattice was approximately at an orientation of 48 degrees away from the compression stress axis (Fig. 4d). The lattice in the surrounding region was misoriented from the 48 degree orientation by approximately, 8 degrees, cf. the 15 degree contour. The maximum temperature was 690^0 C at a region bounded by the free surface of the void and the $X_2 = 0$ boundary (Fig. 4e). The highest value of the mean pressure (all the stresses were non-dimensionalized by Young's modulus) occurred ahead of the free surface of the void on the $X_1 = 0$ boundary, and was approximately .07 of Young's modulus of the single copper crystal (Fig. 4f). These high values of pressure signify that the stress field in this region was highly triaxial.

The contours for the D48 model indicate that an increase in the flow stress, resulted in a decrease of the plastic strains. Geometrical and thermal softening of the crystal was also associated with this increase in the flow stresses at the tip of the void. The slip deformation was constrained to a narrow region bounded by the $X_2 = 0$ boundary and at the tip of the collapsed void. The profiles for the normal stresses, the hoop stresses, and the mean pressures, corresponding to a nominal strain of 8%, are displayed in Figs. 5(a - c). As the stress profiles along the $X_2 = 0.0$ show, there was a region of *unloading* from compression to tension ahead of the tip of the collapsed void. It is in this region of local unloading that the material had also attained maximum values of thermal and geometrical softening. Beyond this region of unloading, the normal stresses had a stress concentration factor of 3.39 (Fig. 5a), the hoop stresses had a stress concentration factor of 3.20 (Fig. 5b), and the mean stresses had a stress concentration factor of 3.53 (Fig. 5c). At a nominal strain-rate of 1000/s and a nominal strain of approximately 9%, the unloaded region had a normalized length of .05 (Fig. 5d) where R is the current deformed radius of the void, and L_1 is the current length of the specimen ahead of the void tip along the $X_2 = 0$ boundary. This compares favorably with the initial crack measurements of Nemat-Nasser and Chang (1990). They reported tensile cracks initially forming at a nominal strain of 7.3%, with an initial normalized crack length of .045, at a nominal strain-rate of approximately 1100/s.

To gain further insight in the micromechanical behavior in the neighborhood of the collapsed void, the Orowan model of plastic flow of single crystals and the viscous drag model are employed (Hirth and Lothe, 1982). The use of the Orowan model can be justified in this region of local unloading because, as the deformation contours show, the primary slip-system strongly dominated the slip process. The Orowan model for single slip is given by

$$\dot{\gamma}^P = b \, N \, v_d, \qquad (28)$$

and the viscous drag model for lattice dislocation motion is given by

$$\tau \, b = \frac{B \, v_d}{(1 - \frac{v_d^2}{c_s^2})^{\frac{1}{2}}}, \qquad (29)$$

where v_d is the dislocation velocity, B is the viscous drag coefficient, b is the modulus of the Burgers vector, N is the dislocation density, τ is the resolved shear stress, c_s is the shear wave velocity, and $\dot{\gamma}^P$ is the plastic slip strain-rate. Eliminating the dislocation velocity between the two above equations results in an equation for the dislocation density,

$$N = \frac{\left[1 + \frac{b^2 \, \tau^2}{c_s^2 \, B^2} \right]^{\frac{1}{2}}}{\tau \, b^2}. \qquad (30)$$

A value of 5.35×10^{-5} Pa-s was used for the viscous drag coefficient and a value of 3.00×10^{-10} m was used for the modulus of the Burgers vector (Vreeland, 1968). Once the dislocation densities are obtained, the dislocation velocity can be calculated from (28). From (29), it is seen that there are two stress regimes; a low stress regime $\tau \ll \frac{Bc_s}{b}$ and a high stress regime where $\tau \gg \frac{Bc_s}{b}$ in which the strain-rate reaches its limiting value. Using values obtained from the finite-element calculations for the maximum plastic primary slip strain-rate and the maximum resolved shear stresses, the following dislocation densities and dislocation velocities were obtained:

Table 2
Dislocation calculations for different hardening models

Hardening Model	$N \times 10^{10}$ $(\frac{1}{m^2})$	$\frac{v_d}{c_s}$	$\frac{\tau_{max}}{G}$	$\dot{\gamma}^p \times 10^{-3}$ $(\frac{1}{s})$	$\frac{T}{T_{melt}}$
D8	4.46	.923	.022	26.9	.45
D48	6.43	.998	.160	42.0	.60
D100	0.78	.995	.088	5.1	.60

The high slip-rates, for the D8 and the D48 model, indicate that the lattice motion was controlled by drag dislocation (Table 2). Furthermore, the calculations clearly show that the plastic slip strain-rates were approaching the limiting value of the elastic shear wave speed for all three hardening models. The theoretical strength of f.c.c. crystals is approximately $\frac{\tau_{max}}{G} = .06$ at a temperature of 630^0 C (Hirth and Lothe, 1982). The theoretical strength is regarded as the threshold stress for the homogeneous nucleation of dislocations. In addition, the elastic shear wave speed is the limiting value of the dislocation velocity. The hardening models of D48 and D100 exceeded this theoretical stress threshold. Table 2 also shows that the temperatures were approaching the recrystallization limit of .65 of the melting temperature (Mecking and Gottstein, 1982). From the calculations in Table 2, for the D48 and the D100 models, it is seen that the limiting values of the primary plastic slip strain-rates and the primary resolved shear stresses can result in the nucleation of dislocations near the tip of the collapsed void (Hirth and Lothe, 1982). The presence of large values of stress concentration and hydrostatic pressure, in combination with the opening mode of stress at the tip of the collapsed void, and the homogeneous generation of dislocations, renders this region of stress unloading a likely site for the formation of cracks. The physically limiting values of the plastic slip strain-rates, the resolved shear stress, and the temperature occurred at compressive nominal strains of approximately 9% for the D48 model. Nemat-Nasser and Chang (1990), reported that cracks initially form at a nominal strain of 7.3%. The location of the unloaded region, in the present study, was ahead of the tip of the collapsed void along the $X_2 = 0$ boundary. As noted earlier, this region of unloading corresponded to the region where the cracks in the experimental study of Nemat-Nasser and Chang (1990) initially form (Fig. 6d).

4.1 Strain-Rate History
To investigate the effects of strain-rate history on compression-induced void collapse, the single crystal was subjected to nominal strain-rates ranging from 200/s to 5000/s. The D48 reference model (Table 1) was used. As the stress-strain curves show (Fig. 6e), the single copper crystal strain-rate hardened; i.e., the load carrying capacity of the monocrystalline material increased as the nominal strain-rate increased. At a nominal strain-rate of 200/s, there was a global unloading of the crystal, however at the same nominal strain, the stresses were still hardening for the 5000/s nominal strain-rate case.

An increase in the nominal strain-rate resulted in an increase in the load carrying capacity of the specimen, and an increase in the resistance of the void to collapse. This has also been experimentally observed by Nemat-Nasser and Chang (1990). The finite-element calculations show that as the nominal strain-rate increased, the formation of localized patterns were retarded. The thermal softening of the crystal and the plastic slip strain-rates increased with increasing applied nominal strain-rates. The dislocation velocities and the dislocation densities for different nominal strain-rates for the D48 model, based on equations (28)-(29), are shown in Table 3. The dislocation densities and the dislocation velocities increased with increasing values of the nominal strain-rate. An increase in the nominal strain-rate resulted in the computed strength of the crystal exceeding the theoretical strength of the crystal (Table 3). Furthermore, temperatures, for the nominal strain-rate of 5000/s for the D48 model, were as high as .60 of the melting temperature of copper. Hence, based on the limiting values of the plastic slip strain-rates, the temperatures, and the theoretical strength of the single crystal, the probability of nucleation of homogeneous dislocations can increase as the nominal strain-rate increases. As with the strain-hardening calculations, there was also a region of local unloading ahead of the collapsing void. This local unloading from compression to tension only occurred for the nominal strain-rates that did not globally unload (Fig. 6e). This region of unloading, in combination with the high values of stress concentration and hydrostatic pressure near the tip of the collapsed void, indicates that an increase in the nominal strain-rate may increase the probability of tensile crack formation in monocrystalline copper.

Table 3
Dislocation calculations for different nominal strain-rates

Nominal Strain-rate $(\frac{1}{s})$	$N \times 10^{10}$ $(\frac{1}{m^2})$	$\frac{v_d}{c_s}$	$\frac{\tau_{max}}{G}$	$\dot{\gamma}^p \times 10^{-3}$ $(\frac{1}{s})$	$\frac{T}{T_{melt}}$
200	.545	.982	.048	3.5	.34
1000	6.43	.998	.160	42.0	.60
5000	17.5	.999	.450	114.2	.60

5. DISCUSSION AND SUMMARY

The high strain-rate mechanisms characterizing the mode of brittle and ductile material failure were investigated for cubic metallic single crystals. It was shown that cleavage fracture in cubic single crystals can occur, and that *tensile* cracks can form at the tip of circular voids that are subjected to compressive nominal strain-rates. Increasing values of the reference stress and the nominal strain-rates resulted in higher flow stresses near the tip of the collapsing void. Large thermal gradients, with temperatures approaching the recrystallization limit near the collapsed void, resulted in the unloading of the flow stresses ahead of the void tip. This local unloading, from compression to tension ahead of the tip of the collapsed void, occurred as the global nominal loads continued to harden. Beyond this region of local unloading, there were large concentrations of the hoop, normal, and mean macroscopic stresses. This increase in the flow stresses was also accompanied by the thermal and geometrical softening of the crystal, which resulted in the local unloading of the flow stresses near the tip of the collapsed void. Furthermore, in this unloaded region, limiting values of dislocation velocity, plastic strain-rates, and the theoretical crystalline strength were attained. These physically limiting values can result in the homogeneous generation of dislocations near the tip of the void. Based on the reference stress that matched the experimental stress-strain response reported by Nemat-Nasser and Chang (1990), this homogenous generation of dislocations in combination with the opening mode of stress and the stress concentrations at the tip of the collapsed void can result in the formation of tensile cracks normal to the compression axis. As noted earlier, these limiting values occurred at approximately the same nominal strain and at the same region as reported by Nemat-Nasser and Chang (1990). Experimental observations by Mughrabi (1983, 1987) indicate that if there are large stress concentrations in single copper crystals subjected to macroscopic tensile loads, cracks will nucleate by dislocation flows in the lattice, and hence *cleavage* failure can occur in normally ductile crystals. Furthermore et al. (1985), Lynch (1986) and Li et al. (1989) have reported experimental results of cleavage-like fracture in some f.c.c. solids, including normally ductile copper. Basinski and Basinki (1985), Jackson (1989), Kwon et al. (1989), and Ma and Laird (1989) have also proposed crack nucleation models for cubic metals, based on dislocation nucleation and pileups, at stress concentrators, grain boundaries, persistent slip bands, and surface asperities. The cleavage fracture of f.c.c. metals, reported in these studies, occurs when the theoretical strength of the crystal has been exceeded.

At lower values of the reference stress and the nominal strain-rates, the global nominal stresses unloaded with increasing values of the nominal compressive strains. This global unloading resulted in the *ductile* failure of crystalline dynamic shear-strain localization. The global unloading was due to the coupled effects of the thermal and geometrical softening of the crystal surmounting the strain hardening and the strain-rate hardening of the material. Geometrical softening resulted in the lattice rotating towards a 45 degree orientation away from the compression stress axis. At this orientation, the resolved shear stresses achieved their maximum values; and this subsequently resulted in an increase in the plastic slip strains and the plastic slip-rates in localized regions of the crystal. The shear band emanated from the collapsed void at a 45 degree orientation, and hence was also oriented approximately 45 degrees from the compression stress axis. The 45 degree orientation is coincident with the maximum macroscopic shear stress

208

orientation, and the computed shear band orientation is consistent with experimental observations on the formation of shear bands in single f.c.c. crystals that are compressively deformed (Hatherly and Malin, 1979). At the lowest value of the reference stress, at a nominal strain-rate of 1000/s, the primary slip system was nearly aligned with the shear band. In this case, the high temperatures in the band, which were approximately .45 of the melting temperature of copper, accelerated the shear-strain localization of the single crystal. Experimental studies; see for example, (Rogers, 1983; Marchand and Duffy, 1988; Giovanola, 1988) have shown that material thermal softening is the primary triggering mechanism for the formation of adiabatic shear bands in metals that have been deformed at high rates of strain.

In summary, the competing material effects of strain hardening, and strain-rate hardening with the thermal and geometrical softening mechanisms of the crystal characterize the mode of failure. For low values of the reference stress and the nominal strain-rate, the geometrical and thermal softening of the crystal resulted in shear-strain-localization. Increasing values of the reference stress and the nominal strain-rate resulted in regions of homogeneous dislocation generation. This generation of dislocations in combination with the large stress concentrations adjacent to the region of local stress unloading, can lead to the formation of tensile cracks normal to the compression stress axis.

Acknowledgment
This work was supported by the National Science Foundation grant # MSS-9110280. The computations were performed on the CRAY Y-MP at the North Carolina Supercomputing Center.

REFERENCES

Butcher, B.M., 1973, "Dynamic Response of Partially Compacted Porous Aluminum During Unloading," *Journal of Applied Physics*, Vol. 44, pp. 4577-3875.

Butcher, B.M., Carroll, M.M. and Holt, A.C., 1974, "Shock-Wave Compaction of Porous Materials," *Journal of Applied Physics*, Vol. 45, pp. 3864-3875.

Carroll, M.M. and Holt, A.C., 1971, "Static and Dynamic Pore-Collapse Relations for Ductile Porous Materials," *Journal of Applied Physics*, Vol. 43, pp. 1626-1635.

Culver, R.S., 1973, "Thermal Instability in Dynamic Plastic Deformation," *Metallurgical Effects at High Strain-Rates*, R.W. Rohde, B.M. Butcher, J.R. Holland and C.H. Kames, eds., Plenum Press, New York, NY, pp. 519-530.

Follansbee, P.S., Ragazzoni, G. and Kocks, U.F., 1984, "The Transition to Drag-Controlled Deformation in Copper at High Strain-Rates," *Mechanical Properties of Materials at High Rates of Strain*, Inst. Phys. Conf. Series, No. 70, pp. 71-80.

Frost, H.J. and Ashby, M.F., 1982, *Deformation Mechanism-Maps*, Pergammon Press, Oxford, pp. 45-55.

Gear, C.W., 1971, "The Automatic Integration of Ordinary Differential Equations," *Communications of the Association of Computing Machinery*, Vol. 14, 176-191.

Giovanola, J.H., 1988, "Adiabatic Shear Banding Under Pure Shear Loading, Part I: Direct Observation of Strain Localization and Energy Dissipation Measurements," *Mechanics of Materials*, Vol. 7, pp. 59-71.

Hatherely, M. and Malin, A.S., 1984, "Shear Bands in Deformed Metals," *Scripta Metallurgica*, Vol. 18, pp. 449-454.

Hirth, J. P. and Lothe, J., 1982, *Theory of Dislocations*, John Wiley and Sons, New York, pp. 120-135.

Hori, M. and Nemat-Nasser, S., 1988, "Mechanics of Void Growth and Void Collapse in Crystals," *Mechanics of Materials*, Vol. 7, pp. 1-13.

Johnson, J.N., 1981, "Dynamic Fracture and Spallation in Ductile Materials," *Journal of Applied Physics,* Vol. 52, pp. 2812-2825.

Klopp, R.W., Clifton, R J. and Shawki, T.G., 1985, "Pressure-Shear Impact and the Dynamic Viscoplastic Response of Metals," *Mechanics of Materials,* Vol. 4, pp. 375-385.

Marchand, A. and Duffy, J., 1988, "An Experimental Study of the Formation Process of Adiabatic Shear Bands in Structural Steel," *Mechanics and Physics of Solids,* Vol. 36, pp. 251-283.

Mecking, H. and Gottstein, G., 1978, "Recovery and Recrystallization During Deformation," *Recrystallization of Metallic Materials,* F. Haessner, eds., Reiderer-Verlag, Stuttgart, pp. 195-215.

Mughrabi, H., 1983, "Dislocation Wall and Cell Structures and Long Range Internal Stresses in Deformed Metal Crystals," *Acta Metallurgica,* Vol. 31, pp. 1367-1379.

Mughrabi, H., 1987, "A Two Parameter Description of Heterogeneous Dislocation Distributions in Deformed Metal Crystals," *Materials Science and Engineering,* Vol. 85, pp. 15-31.

Nemat-Nasser, S., and Chang, S., 1990, "Compression-Induced High Strain-Rate Void Collapse, Tensile Cracking, and Recrystallization in Ductile Single Crystals and Polycrystals," *Mechanics of Materials,* Vol. 10, pp. 1-17.

Rogers, H. C., 1983, "Adiabatic Shearing-General Nature and Material Aspects," *Material Behavior under High Stress and Ultrahigh Loading Rates* , J. Mescall and V. Weiss, eds., Plenum Press, New York, pp. 101-118.

Vreeland, Jr., T., 1968, "Dislocation Velocity in Copper and Zinc" *Dislocation Dynamics,* A.R. Rosenfeld, G.T. Hahn, A.L. Bement, and R.I. Jaffee, eds., McGraw Hill, New York, pp. 529-546.

Zikry, M. A., and Nemat-Nasser, S., 1990, "High Strain-Rate Localization and Failure of Crystalline Materials," *Mechanics of Materials,* Vol. 10, pp. 215-237.

Zikry, M.A., 1993a, "Dynamic Void Collapse And Material Failure Mechanisms In Metallic Crystals," *Mechanics of Materials,* in press.

Zikry, M.A., 1993b, "An Accurate And Stable Algorithm For High Strain-Rate Finite Strain Plasticity," *Computers and Structures,* in press.

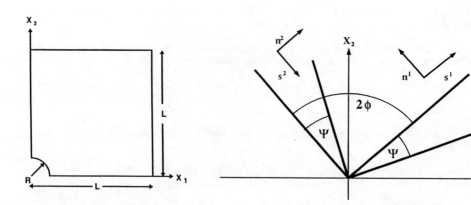

Fig. 1a Plate and void geometry.   Fig. 1b Double slip model.

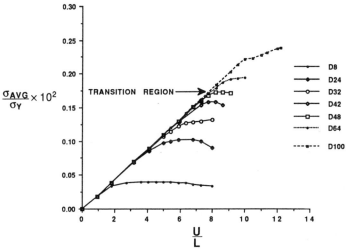

Fig. 1c   Stress-strain curve for different reference stress models.

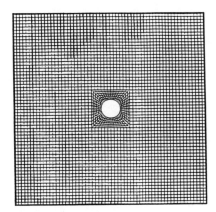

Fig. 2a Undeformed mesh

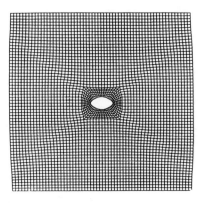

Fig. 2b Deformed mesh for the D8 model at
5% nominal strain.

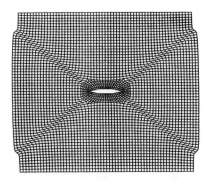

Fig. 2c   Deformed mesh for the D8 Model at 8% nominal strain.

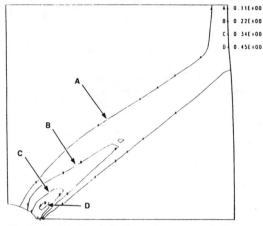

A = 0.11E+00
B = 0.22E+00
C = 0.34E+00
D = 0.45E+00

Fig. 3a Accumulated plastic strains at a nominal strain of 5% (D8 Model).

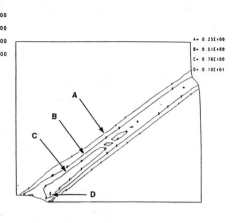

A = 0.25E+00
B = 0.51E+00
C = 0.76E+00
D = 0.10E+01

Fig. 3b Accumulated plastic strains at a nominal strain of 8% (D8 Model).

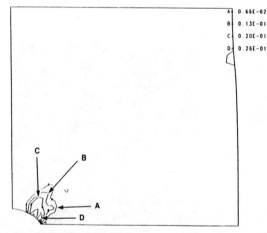

A = 0.66E-02
B = 0.13E-01
C = 0.20E-01
D = 0.26E-01

Fig. 3c Primary slip-strains at a nominal strain of 5% (D8 Model).

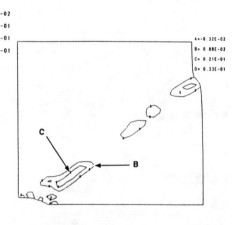

A = -0.32E-02
B = 0.88E-02
C = 0.21E-01
D = 0.33E-01

Fig. 3d Primary slip-strains at a nominal strain of 8% (D8 Model).

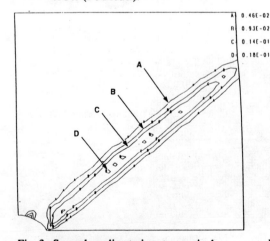

A = 0.46E-02
B = 0.93E-02
C = 0.14E-01
D = 0.18E-01

Fig. 3e Secondary slip-strains at a nominal strain of 5% (D8 Model).

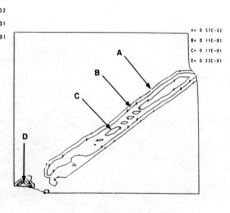

A = 0.57E-02
B = 0.11E-01
C = 0.17E-01
D = 0.23E-01

Fig. 3f Secondary slip-strains at a nominal strain of 8% (D8 Model).

212

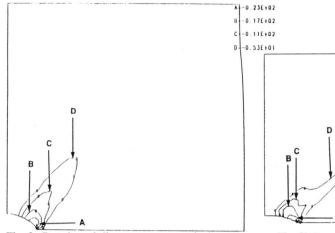

A - 0.23E+02
B - 0.17E+02
C - 0.11E+02
D - 0.53E+01

Fig. 3g Rotation of slip-system at a nominal strain of 5% (D8 Model).

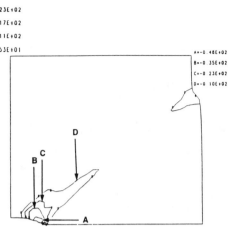

A=-0.48E+02
B=-0.35E+02
C=-0.23E+02
D=-0.10E+02

Fig. 3h Rotation of slip-system at a nominal strain of 8% (D8 Model).

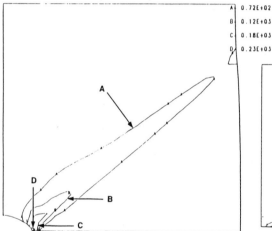

A 0.72E+02
B 0.12E+03
C 0.18E+03
D 0.23E+03

Fig. 3i Temperature distribution at a nominal strain of 5% (D8 Model).

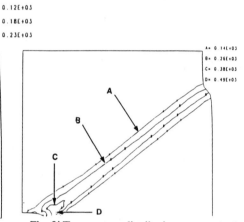

A= 0.14E+03
B= 0.26E+03
C= 0.38E+03
D= 0.49E+03

Fig. 3j Temperature distribution at a nominal strain of 8% (D8 Model).

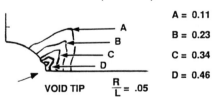

A = 0.11
B = 0.23
C = 0.34
D = 0.46

VOID TIP $\frac{R}{L}$ = .05

Fig. 4a Accumulated plastic strains at a nominal strain of 9% (D48 Model).

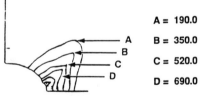

A = 190.0
B = 350.0
C = 520.0
D = 690.0

Fig. 4b Primary slip-strains at a nominal strain of 9% (D48 Model).

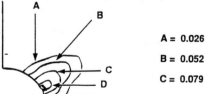

A = 0.026
B = 0.052
C = 0.079

Fig. 4c Secondary slip-strains at a nominal strain of 9% (D48 Model)

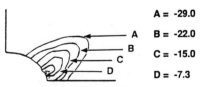

A = -29.0
B = -22.0
C = -15.0
D = -7.3

Fig. 4d Rotation of slip-system at a nominal strain of 9% (D48 Model).

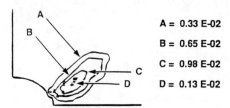

A = 0.33 E-02

B = 0.65 E-02

C = 0.98 E-02

D = 0.13 E-02

Fig. 4e Temperature distribution at a nominal
strain of 9% (D48 Model).

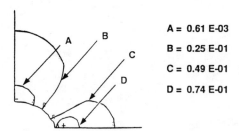

A = 0.61 E-03

B = 0.25 E-01

C = 0.49 E-01

D = 0.74 E-01

Fig. 4f Pressure distribution at a nominal strain
of 9% (D48 Model).

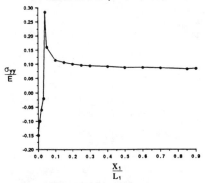

Fig. 5a Normal stress concentration
(D48 Model).

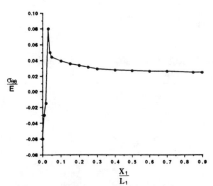

Fig. 5b Hoop stress concentration
(D48 Model).

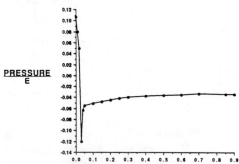

Fig. 5c Mean stress concentration
(D48 Model).

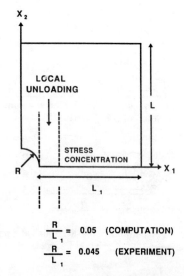

$$\frac{R}{L_1} = 0.05 \quad \text{(COMPUTATION)}$$

$$\frac{R}{L_1} = 0.045 \quad \text{(EXPERIMENT)}$$

Fig. 5d Region of local stress unloading and
stress concentration near the void.

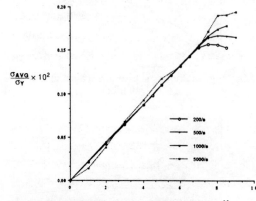

Fig. 5e Stress-strain curves for different
nominal strain-rates (D48 Model).

214

AD-Vol. 36, Fatigue and Fracture of Aerospace Structural Materials
ASME 1993

ON TIME DEPENDENT DEFORMATION MODELING
OF METAL MATRIX COMPOSITES

J. Ahmad, U. Santhosh, and I. U. Haq
AdTech Systems Research Incorporated
Dayton, Ohio

ABSTRACT

A simple engineering model is presented to predict time dependent response of unidirectional metal matrix composites (MMCs) under sustained loading in the fiber direction. The model takes into account the effect of matrix creep and consolidation process induced residual stresses. Model predictions are compared with experimental data on SCS-6/Ti-6Al-4V composite and with nonlinear finite element analyses. Also, a simple modeling approach is presented to assess the effect of possible fiber fractures on the composites global strain-time response.

1. INTRODUCTION

Titanium alloy and intermetallic matrix composites (TMCs) with continuous ceramic fibers are candidate materials for several advanced airframe and turbine engine components. In many of these applications, temperatures and stresses are sufficiently high to cause inelastic deformation of the matrix material which contributes, in part, to inelastic deformation of the composite. Other contributions to the composite's global inelastic deformation response come from microscopic damage mechanisms such as reaction zone and matrix microcracking, fiber-matrix debonding and localized fiber fractures.

In the present paper we focus attention on time dependent inelastic deformation response of MMCs under constant applied load in the fiber direction. The temperature and applied loads considered are such that the matrix material (Ti-6Al-4V) is expected to creep upon initial application of the load. The silicon carbide fibers (SCS-6) are expected to remain linearly elastic to failure. Under these conditions, and knowing the mechanical behavior of the fiber and the matrix, we attempt to predict the composite's deformation response as a function of time. Similar such attempts have been reported in the literature by Schwenker et.al. [1] and by Wright [2].

Using a simple one dimensional model proposed by McLean[3], Schwenker et.al. attempted to predict unidirectional SCS-6/Ti-6Al-4V composite response assuming the matrix behavior to be elastic-secondary creep. Thus, the matrix creep strain rate was represented by the following equation:

$$\dot{\varepsilon}_m = B_m \, \sigma_m^n \qquad\qquad (1)$$

where B_m and n are temperature dependent steady state creep constants for the matrix material obtained by performing a number of creep tests on the matrix material at several temperatures and stress levels. The fiber material was assumed to be isotropic and linear elastic. Tables 1 and 2 give the fiber and matrix Young's moduli (E) and Poisson's ratios (ν) at two temperatures, with subscripts m and f designating matrix and fiber, respectively. Table 2 also gives the matrix material creep constants at the two temperatures. The values of elastic moduli and creep constants are from reference [1]. The Poisson's ration (ν) and coefficient of thermal expansion (α) values for the two materials (also shown in Tables 1 and 2 but not used in reference [1]), are from reference [4].

Table-1: Elastic Properties of SCS-6 [1,4]

Temp. (°C)	E_f (GPa)	ν_f	α_f (μm/m/°C)
427	374	0.25	3.90
538	354	0.25	4.07

Table-2: Creep Properties of Ti-6Al-4V [1,4]

Temp. (°C)	E_m (GPa)	ν_m	n	B_m (MPa^{-n}s^{-1})	α_m (μm/m/°C)
427	85	0.30	8	2.58×10^{-29}	9.81
538	69	0.30	7	2.12×10^{-23}	9.77

Schwenker et.al. found that for the SCS-6/Ti-6Al-4V composite with fiber volume fraction (V_f) of 0.35, the model proposed by McLean consistently under-estimated the experimentally measured strains in the composite by large amounts.

Using a "two-bar model", Wright [2] also found that at 649°C, the predicted strain in a SCS-6/Ti-24Al-11Nb composite were considerably lower than those measured experimentally. The two-bar model used in the analysis is similar to that of McLean except that the matrix material creep strain rate is allowed to have both stress and time dependence in the following manner:

216

$$\dot{\varepsilon}_m = B_m \, \sigma_m^n t^m \qquad\qquad (2)$$

Schwenker et.al. [1] have discussed some of the possible reasons for the discrepancy between predicted and measured strains. Their main focus was on deformation caused by micromechanical damage in the form of fiber fractures close to the machined edges of their dog-bone shaped flat creep test specimens tested in air. Their test specimens were 5.0mm wide in the gauge section, 10.0 mm wide in the grip section and 8-ply (1.78 mm) thick. With 0.142 mm diameter SCS-6 fibers and 0.33 to 0.35 fiber volume fraction, the specimen contained an average of 23 fibers across the gauge section width. They observed that at 538°C, one to three fibers were fractured near each edge; either during machining or after prolonged exposure of the specimen to the environment under load. No estimates are given in reference [1] on the amount of additional deformation which can be attributed to the observed fiber breaks. In the present paper a simple model is presented which can be used to assess the effect of fiber fracture on the global time dependent deformation behavior of the composites.

Another source of additional deformation in creep tests may be the presence of consolidation process induced tensile residual stress in the matrix material. This aspect was not included in the models used in references [1] and [2], but was recently addressed by Ahmad et.al.[5]. In the following, a summary of the reference [5] approach is presented as a prologue to the model for assessing the effect of fiber breaks.

2. MODELING OF COMPOSITE DEFORMATION DUE TO MATRIX CREEP

Assume the fiber material to be linearly elastic and the matrix to behave in an elastic-secondary creep manner, with creep strain rate given by equation (1). Also assume that the fibers and matrix remain bonded. Then, simple one dimensional consideration of force equilibrium and strain compatibility under constant zero degree loading of a unidirectional laminate gives the following expression for the rate of strain in the composite:

$$\dot{\varepsilon} = B_u \, \sigma_m^n \qquad , \qquad\qquad (3)$$

$$\text{where} \qquad B_u = \frac{B_m E_m V_m}{E_c} \quad , \qquad\qquad (4)$$

$$E_c = E_f V_f + E_m V_m \ , \qquad\qquad (5)$$

$$V_m = (1 - V_f) \qquad , \qquad\qquad (6)$$

and matrix stress (σ_m) includes the contribution of applied constant stress (σ_c) and the consolidation process induced residual stress (σ_m^T), i.e.

$$\sigma_m = \frac{\sigma_c}{V_m} \left[1 - \frac{E_f V_f}{\sigma_c} \, \varepsilon \right] + \sigma_m^T \qquad\qquad (7)$$

Assuming uniform strain in the fiber and matrix and purely elastic straining during consolidation, an estimate for the matrix residual stress can be expressed as follows:

$$\sigma_m^T = \frac{E_f E_m V_f (\alpha_f - \alpha_m)}{E_c} \Delta T \quad , \tag{8}$$

where ΔT is the temperature change between consolidation temperature and the test temperature and is a negative number.

Integrating equation (2), the time dependent portion of the composite strain can be expressed as follows:

$$\varepsilon^c(t) = \frac{P - y(t)}{Q} - \frac{\sigma_c}{E_c} \quad , \tag{9}$$

$$\text{where,} \qquad P = \left(\frac{B_m E_m V_m}{E_c} \right)^{1/n} \left(\frac{\sigma_c}{V_m} + \sigma_m^T \right) \quad , \tag{10}$$

$$Q = \left(\frac{B_m E_m V_m}{E_c} \right)^{1/n} \cdot \frac{E_f V_f}{V_m} \quad , \tag{11}$$

$$y(t) = \left[(n-1)Qt + \frac{1}{y_0^{n-1}} \right]^{\frac{-1}{n-1}} \quad , \tag{12}$$

$$\text{and} \qquad y_0 = P - (Q\sigma_c/E_c) \, . \tag{13}$$

Figure-1 shows a plot of the composite's time dependent or "creep" strain at 655 MPa constant applied stress using the material constants given in Tables 1 and 2 at 538°C. Also shown in the figure are experimental data from reference [1] and results of nonlinear finite element analyses (FEA) from reference [5]. The finite element analyses were performed using the same power-law creep model for the matrix material as given by equation (1). Both the finite element results and equation (9) solutions (designated as "creep model" in Figure-1) were obtained with and without the consideration of residual stress (σ_m^T). In both cases the simple creep model gives results which are in close agreement with FEA estimates. Also, inclusion of residual stresses significantly improves the agreement between model predictions and the experimental data at 655 MPa applied stress.

Figure-2 shows the results at 965 MPa applied stress including the effect of residual stresses. It is seen that the creep model and FEA results are again in good agreement. However, at this higher stress level, the effect of residual stress falls short of providing a good agreement between predictions and experimental data. In the following we explore this discrepancy by formulating and using a simple model to account for possible contributions to the total composite strain by broken fibers.

3. MODELING OF DEFORMATIONS DUE TO BROKEN FIBERS

Consider a unidirectional laminate of length $2l$ and width $2W$, a symmetric half of which is schematically depicted in Figure-3. Consider that at each edge of the panel there is a region B of width W_B in which all the fibers have fractured along the symmetry line. The remaining width of the panel ($2 W_U$) has no broken fibers. The panel is subjected to uniform applied displacement δ_c at each end, such that the global composite strain is δ_c/l.

Figure-4 depicts the cross-section of a unit cell within the region B. It consists of a region B_1, in which the composite behaves as in the unbroken fiber region U, and a region B_2 which contains the fiber fracture location and its associated fiber-matrix split of length 2a. Ahead of each tip of the fiber-matrix split is a distance 2s over which some load transfer occurs by shear. That is, 2s is a shear lag distance. The shear stress varies from some high value at the tip of the split to zero at a distance 2s from the tip in some fashion. Assuming this variation to be linear, one can say that the fiber carries no normal stress over a distance d on each side of the fiber fracture location, where d is the sum of a and s. Thus, in the region B_2, the matrix material supports the entire applied load on region B.

Denoting stress rate by $\dot{\sigma}$, with subscripts U, B, B_1 and B_2 indicating the regions shown in Figures 3 and 4, we have

$$\dot{\sigma}_{B_1} = \frac{W}{W_B}\left(\dot{\sigma}_c - \frac{W_u}{W}\dot{\sigma}_u\right) \tag{14}$$

and $\qquad \dot{\varepsilon} = \dot{\varepsilon}_B = \dot{\varepsilon}_u \tag{15}$

form equilibrium and compatibility considerations.

For the region U with no broken fibers, the creep model from the previous section can be used to write.

$$\dot{\varepsilon}_u = \frac{\dot{\sigma}_u}{E_c} + B_u \sigma_{mu}^n \tag{16}$$

where σ_{mu} denotes matrix stress in the region U. For region B,

$$\dot{\varepsilon}_B = \frac{\dot{\sigma}_{B_1}}{E_B} + \frac{l-d}{l}B_u\sigma_{mB_1}^n + \frac{d}{l}B_m\left(\frac{\sigma_{B_1}}{1-V_f}\right)^n \ , \tag{17}$$

where σ_{mB_1} denotes matrix stress in the region B_1 and

$$E_B = \frac{lV_mE_mE_c}{dE_c + (l-d)E_mV_m} \tag{18}$$

Equations (14) thru (18), together with the assumption that the fibers remain elastic, lead to the following expression for the composite's global strain rate ($\dot{\varepsilon}$):

219

$$\left(1 + \frac{W_u E_c}{W_B E_B}\right) \dot{\varepsilon} = \frac{W_u E_c}{W_B E_B} B_u \left[\frac{\sigma_u}{V_m} - \frac{E_f V_f}{V_m} \varepsilon + \sigma_m^T\right]^n$$
$$+ \left(1 - \frac{d}{l}\right) B_u \sigma_{mB_1}^n + \frac{d}{l} B_u \left[\frac{W}{W_B V_m}\left(\sigma_c - \frac{W_u \sigma_u}{W}\right)\right]^n \qquad (19)$$

where,

$$\sigma_{mB_1} = \frac{W}{W_B V_m}\left(\sigma_o - \frac{W_u}{W}\sigma_u\right) - \frac{E_f V_f}{V_m}\left(\frac{l}{l-d}\right)\left(\varepsilon - \frac{d}{l}\varepsilon_{B_2}\right) + \sigma_M^T \, , \qquad (20)$$

and σ_u and ε_{B_2} can be found by integrating the following expressions:

$$\dot{\sigma}_u = E_c\left[\dot{\varepsilon} - B_u \left(\frac{\sigma_u}{V_m} - \frac{E_f V_f}{V_m}\varepsilon + \sigma_m^T\right)^n\right] \qquad (21)$$

$$\dot{\varepsilon}_{B_2} = B_m\left[\frac{W}{W_B V_m}\left(\sigma_o - \frac{W_u}{W}\sigma_u\right)\right]^n \qquad (22)$$

The initial conditions to be used in evaluating σ_u and ε_{B_2} are:

$$\sigma_u(t = 0) = \left(\frac{W}{W_u} - \frac{W_B E_B}{W_u E_o}\right)\sigma_c \qquad (23)$$

and
$$\varepsilon_{B_2}(t = 0) = \frac{\sigma_c E_B}{V_m E_m E_o} \qquad (24)$$

where
$$E_o = E_c \frac{W_u}{W} + E_B \frac{W_B}{W} \qquad (25)$$

The total strain in the composite panel is found by numerically integrating equation (19). In the present work, this was done by using the fourth order Runge-Kutta method. The initial elastic strain of the panel can be expressed as:

$$\varepsilon(t = 0) = \sigma_c/E_o \qquad (26)$$

Figure-5 shows the predicted time dependent or "creep" strain in the composite panel for W_B/W of 0.1 and various d/l ratios. The results correspond to 538°C and applied stress of 965 MPa. It is seen that creep strain changes by less than ten percent for four orders of magnitude change in d/l. thus, the solution is not very sensitive to the choice of d/l.

In Figure-6, results are presented for $d/l = 0.1$ percent and the width of the region with broken fibers ranging from 0.1 to 30 percent of the total width of the panel. Comparing with the experimental curve in Figure-6 the theoretical curve agrees with the former for the case $W_B/W = 0.25$, corresponding to approximately three rows of broken fibers at each specimen edge. This is in qualitative agreement with experimental observations reported in reference [1] for tests at 538°C.

4. CONCLUSIONS

A simple modeling approach was adopted to predict time dependent deformation response of unidirectional MMC laminates subjected to constant loads along the fibers. It was found that to get reasonably good agreement between predictions and experimental data, one should include the consideration of consolidation process induced residual stresses in the model.

At higher temperatures and stress levels, it may be necessary to include the consideration of fractured fibers in the analysis models to account for additional global strain in the composite which cannot be attributed to matrix material creep. A simple model has been derived which includes the consideration of broken fibers on the global strain. The model, in conjunction with a criterion for fiber breakage, can be used in predicting global strain response with increased accuracy.

REFERENCES

1. Schwenker, S.W., Roman, I., and Eylon, D., 'Creep Behavior of SCS-6Ti-6Al-4V Unidirectional Composites', in the Proceedings of the International Conference on Advanced Composites 1993, University of Woolongong, Australia, Feb. 15-19, 1993.

2. Wright, P.K., 'Creep Behavior and Modeling of SCS-5/Titanium MMC', in Titanium Matrix Composites Workshop Proceedings, P.R. Smith and W.C. Revelos, Ed., Technical Report WL-TR-92-4035, Wright Laboratory, Wright-Patterson AFB, Ohio, 1992, pp. 251-276.

3. McLean, M., 'Mechanisms and Models of High Temperature Deformation of Composites', Materials and Engineering Design: The Next Decade, Eds. B.F. Dyson and D.R. Hayhurst, Inst. of Metals, London, 1989, pp. 287-294.

4. Nimmer, R.P., Bankert, R.J. , Russell, E.S. , Smith, G.A. and Wright, P.K., 'Micromechanical Modeling of Fiber/Matrix Interface Effects in Transversely Loaded SiC/Ti-6-4 Metal Matrix Composites', *Journal of Composites Technology and Research*, Vol. 13, No. 1, Spring 1991, pp. 3-13.

5. Ahmad, J., Santhosh, U. and Haq, I.U., 'An Analysis of Time Dependent Deformation of Metal Matrix Composites', to appear in Proceedings of the 8th Technical Conference on Composite Materials, American Society for Composites, October 1993.

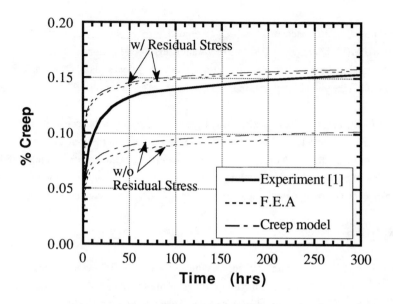

Figure-1: Creep of 33% SCS-6/Ti-6-4 lamina at 538°C and 655 MPa

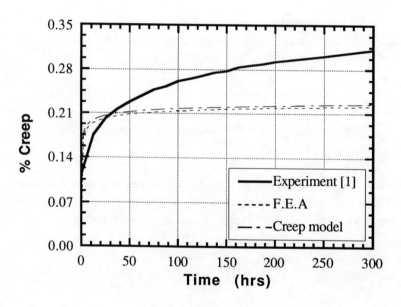

Figure-2: Creep of 33% SCS-6/Ti-6-4 lamina at 538°C and 965 MPa

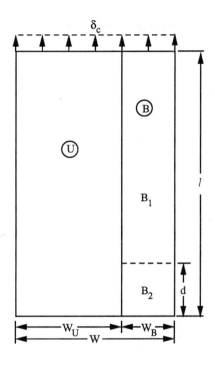

Figure-3: Upper symmetric half of a composite panel with broken fibers in region B

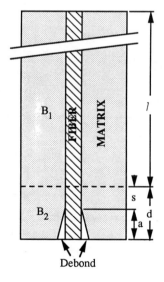

Figure-4: Cross-section of a unit cell within the broken region B of Figure-3

223

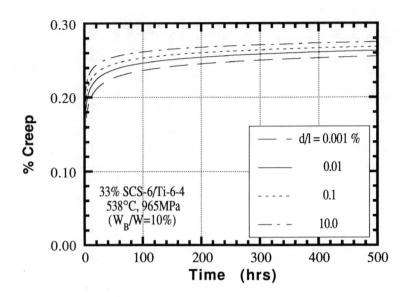

Figure-5: Effect of fiber debond length on composite creep

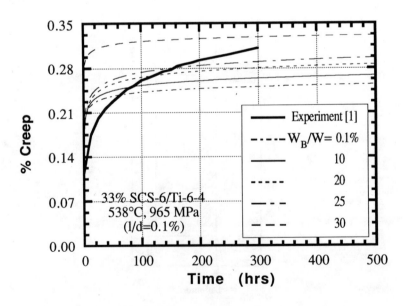

Figure-6: Effect of volume of broken fibers on composite creep

AUTHOR INDEX

Fatigue and Fracture of Aerospace Structural Materials